Three
Second Edition

Service improvement using value stream design

Simon Dodds

www.ThreeWins.com

First published in 2007.

© 2006, 2007 by Simon Dodds.

The right of Simon Dodds to be identified as the author of this work has been asserted in accordance with the Copyright, Designs and Patents Act 1988.

ISBN: 978-1-84753-631-0

British Library Cataloguing in Publication Data
A catalogue record for this book is available from the British Library.

All rights reserved. No part of this book may be reprinted or reproduced or utilised in any form or by any electronic, mechanical, or other means, now known or later invented, including photocopying and recording, or in any information storage or retrieval system, without the permission in writing from the publishers. This book may not be lent, resold, hired out or otherwise disposed of by way of trade in any form, binding or cover other than that in which it is published, without the prior consent of the publishers.

The first edition of this book, by Simon Dodds, was published in 2006 as *Three Wins: Service redesign through flow modelling* by Kingsham Press, Chichester, UK. (ISBN 1-904235-54-9).

www.ThreeWins.com

Foreword to the First Edition

In the summer of 1999 I started a new job as a consultant surgeon in a newly formed department of Vascular Surgery at Good Hope Hospital, a medium sized district general hospital on the north east edge of Birmingham. Like all new consultants I was full of enthusiasm and I had lots of ideas about how I would like to change some of the things I had seen during my training. Five years later I had the honour of accepting, on behalf of the whole team, the first NHS Innovation Award for Service Delivery for the re-designed Vascular Surgery Outpatient Clinic and Leg Ulcer Telemedicine Service. An unexpected outcome of that day was an invitation to write about what we did to achieve this award and how we achieved the elusive win-win-win outcome: a better service to patients; a skilled, motivated and enthusiastic team; and a substantial cost saving in treatment costs. We certainly did not anticipate the far-reaching implications of the project when we started; or the enthusiastic support we would get from patients and community NHS staff; or the many visitors that have taken the time to visit and to share knowledge and experience. We had some luck and we had a lot of encouragement but I don't think they were the recipe for success. We had no steering committee, no project plan, no budget, no meetings and no commercial support and I don't think that was the recipe for success either. We put the needs of the patient first; we used the ideas, skills and enthusiasm of the staff; we assessed each incremental change methodically and scientifically at each stage; and we did not give up. Was that the recipe for success?

When I was asked to tell our story I realised that I couldn't describe how we did it - so I couldn't pass on this knowledge. Since then I have done a lot of reading about innovation, how change happens in large organisations and how effective teams are needed to deliver effective change. I have seen common threads in these books that resonate with what we did and I have come to realise that we demonstrated a sixth sense that steered us away from major obstacles which would have meant certain failure. I have come to realise that the essential ingredients for success are:

1. A shared passion that constantly drives the search for a solution.
2. An insatiable curiosity and no fear of considering new ideas.
3. The collective skills and experience of a cohesive team that allowed us to leap hurdles, to learn from setbacks and to deliver solutions.

Yes, we needed some resources and were very fortunate to have been awarded a small research grant at the start; an investment has been repaid with interest! We have enjoyed positive outcomes other than a better service and the kudos of a national award; we have gained a better understanding of clinical process redesign and we have developed and tested methods that make this complex process easier, more predictable and less dependent on luck. My experience has deepened my belief that the future of the NHS lies in the hearts and minds of the NHS staff and it is this optimism for the future and my enthusiasm to keep "shaking the trees" that I hope this book will convey. Some might consider that what we have achieved is unique. I do not think so - I believe anyone can do it.

Sutton Coldfield
October 2005.

Foreword to the Second Edition

Almost two years have passed since writing the first edition of *Three Wins* and I have learned more about service improvement in that time than ever before. There is a revolution underway; a transformation of the way we think about work – and it crystallised in Japan in the 1950's in an automobile company called Toyota. The philosophy and principles that Toyota developed and refined over several decades have enabled them to grow to become the largest car manufacturer in the world. This dramatic transformation has been given a label - *Lean Thinking*. When I heard the term late in 2005 and read about it I realised immediately that it was what we had been doing since 1999. We just didn't call it that; we called it *Common Sense*. Further enquiry revealed two other schools of thought in manufacturing that were broadcasting very similar messages: Motorola with their *Six Sigma* and Goldratt with his *Theory of Constraints*. What I found fascinating was that the underlying principles were the same; they all focussed on improving the flow of work by designing the mistakes and delays out of the processes. It is not surprising that this innovation is now diffusing into service industries such as healthcare and my optimism for the future of healthcare remains high because we now have evidence that the principles of value stream improvement do work. It is now just a matter of learning how to do it and putting that knowledge into practice.

Sutton Coldfield
June 2007

About the Author

Simon Dodds was born in Yorkshire in the "swinging sixties" and in 1978 he was awarded an Open Scholarship to read Medicine at Pembroke College, Cambridge. After completing the Part I Medical Science Tripos he chose a Part II in Computer Science; graduating with 1st Class Honours in 1982. He spent that summer in Cambridge working as computer programmer for a biotechnology start-up company; just when the personal computer revolution was starting and the Internet was being invented. In 1985 he completed his clinical studies at St Bartholomew's Hospital, London and went on to train in general surgery in London, Cambridge and the cathedral towns of Wessex. He was elected fellow of the Royal College of Surgeons of England in 1990 and joined the Wessex 'Flyer' Rotation. He devoted two years to research on haemodynamic modelling of occlusive arterial disease, working in collaboration with the Cavendish Laboratory in Cambridge and was awarded his Master of Surgery degree in 1994. He continued his higher surgical training in Vascular Surgery, during which time he developed an interest in the cause and management of leg ulcers. He was appointed as consultant surgeon at Good Hope Hospital in 1999 and since then has combined his training in surgery, research and computer science in the development of the vascular surgery services. He has always been passionate about innovation and in 2004 the vascular surgery outpatient team were awarded the first NHS Innovation Award for Service Delivery for the development, testing and successful implementation of the Leg Ulcer Telemedicine System which is now being used throughout North East Birmingham and at several other sites in the UK. His current interests include the application of discrete event simulation to the design of improved healthcare processes, work for which he was awarded the Health Information Technology Effectiveness Award 2005 for the Best use of Information Technology in the Health Service. Since writing the first edition of *Three Wins* he has continued to develop a better understanding of the psychodynamics of organisational change and to create training courses and tools to support healthcare "value stream improvement". Simon lives a short walk from Good Hope Hospital with wife Abigail, daughters Alice and Sophie, and Juma the black Labrador. He prefers not to offer a private surgical service and has no understanding of, nor ability to play golf.

Acknowledgements

There are dozens of people who have directly and indirectly contributed to the work described in this book and it is impossible to mention everyone by name – you know who you are and I thank you all for your enthusiasm, support and hard work.

There are some individuals that I must mention by name because their contributions have been critical: Sue Hayes for her clinical skills, patience, exceptional diplomacy and sheer dedication to the whole project; Peter Ingham for many long discussions and help in the critical testing phase; all the district nursing teams in North Birmingham who contributed to the LUTM research trial and who convinced their PCT to adopt the service; Val Robson and Robin Cooper for their infectious enthusiasm and help with some inevitable technical glitches; David Gleaves for suggesting that we should submit the work for the NHS Innovation Award; Wendy Smith for her help in implementing the clinic booking templates; Andy Comber for insisting that we submit the work for the HITEA Award; and Hugo Minney for testing the DES software! I must also acknowledge the contribution made by my three mentors: Stephen Gatley, Mike Thompson and Tony Chant, who over the last twenty years have sowed the seeds of many of the ideas that have led to this book. I would particularly like to thank Anand Kumar of Kingsham Press for suggesting I write a book at all; and Trevor Gay, Sharon Palser, and Dave Pothier for their comments on the first drafts.

Since the first edition was published I have had the pleasure and honour to meet, to be inspired by, and to learn from an increasingly large network of amazing people; people who share my passion for doing whatever we can to help those who entrust us with their care; and to improving the world that our children will inherit. I must pay particular tribute to Andy Ferguson who infectious enthusiasm and insatiable desire to help others is an inspiration to everyone.

Last and most important, I would like to thank Abigail, Alice and Sophie for their love, support, patience and understanding during the whole project and during the writing of *Three Wins*. It has cost you the most of anyone.

Contents

Foreword to the First Edition	iii
Foreword to the Second Edition	iv
About the Author	v
Acknowledgements	vi
Contents	vii
Chronological List of Events	ix
Preface	x

Chapter 1. A new beginning	**1**
The context	2
The challenge	4
The outcome	4
The goal	5
The action	7
The asking	8
The listening	9
The team	13
The learning	15
Chapter 2. Win-win-win	**19**
The motivation	20
The plan	22
Chapter 3. Innovation	**25**
The brainstorm	26
The map	27
The luck	29
Chapter 4. Investigation	**31**
The filter	32
The diffusion	34
The integration	38
The research	39
The report	40
The tipping point	43
The adoption	44
Chapter 5. Implementation	**49**
The path	50
The finish	51

Chapter 6. Complexity   **53**
The game   54
The paradigms   55
The constraints   58
The solution   59
The model   60
The simulation   63

Chapter 7. Value Stream Design   **69**

Chapter 8. The Change Engine   **71**
The spiral   72
The lifecycle   72
The test   74

Final thoughts   **75**

References   **77**

Appendix A - The Innovation Questionnaire   **79**

Appendix B – The First Ten Steps   **83**

Chronological List of Events

Aug 1999	Simon Dodds appointed consultant surgeon at Good Hope Hospital.
Sep 1999	Prospective outpatient audit commenced.
Dec 1999	Leg Ulcer Telemedicine (LUTM) study research proposal submitted.
Jan 2000	LUTM study research grant awarded.
June 2000	One Stop Vascular Outpatient Clinic implemented.
July 2000	Wound measurement software development started.
April 2001	Sue Hayes (research nurse) joined the team to coordinate LUTM study.
July 2001	LUTM software development started.
Nov 2001	Technical feasibility of LUTM proven with the help of Peter Ingham.
Jan 2002	First patient recruited to LUTM study.
Mar 2002	Publication of computerised wound measurement method.
Jul 2002	Visits to Good Hope Hospital by Val Robson and Robin Cooper.
Nov 2003	Last patient completed LUTM study.
Jan 2004	LUTM study report completed.
Apr 2004	LUTM implemented by Robin Cooper in North Hampshire.
July 2004	NHS Innovation Award for Innovative Service Delivery, London.
Aug 2004	North Birmingham PCT rollout of LUTM started.
Nov 2004	DES designed clinic booking template implemented.
Mar 2005	HITEA Best Innovative Use of IT.
	HITEA Best use of IT in the Health Service
Apr 2005	West Midlands NHS Innovations Innovative Service Delivery Award.
Apr 2005	East Birmingham PCT rollout of LUTM started.
Apr 2005	Burntwood Lichfield and Tamworth PCT rollout of LUTM started.
Sep 2005	New Treatment Centre opens at Good Hope Hospital.
Jan 2006	First edition of "Three Wins" published.
Jan 2006	First "Three Wins" workshop.
Feb 2006	Second "Three Wins" workshop co-presented with Andy Ferguson.
Jun 2006	First demonstration of Lean methods at Good Hope Hospital.
Sep 2006	First Value Stream Mapping event at Good Hope Hospital.
Nov 2006	First Rapid Improvement event at Good Hope Hospital.
Apr 2007	Good Hope Hospital becomes part of Heart of England Foundation Trust.
Aug 2007	Second edition of "Three Wins" published.

This book is two stories in one.

> The first is the real story of the Leg Ulcer Telemedicine Project; a blow by blow account, warts and all. These are the facts – the what, who, where, when and why.

The second is the story of **how** and is told in parallel with the first but in reality only become clear after the work had been done with the benefit of hindsight and during the writing of this book.

Preface

What is this book about?
This book tells the story of how a small group of healthcare professionals, brought together by chance, were united in their common desire to improve the care they delivered; and how they re-invented the way in which they work to create a better service for their patients and a better environment for themselves. It describes how, by recognising and using the diverse experiences and skills of every member of the team; by persistence and focusing on the things that they valued most; and by using ideas and techniques borrowed from elsewhere, their clinical service was gradually, in relatively easy steps, evolved into probably the best service of its kind in the UK, and possibly the world. This book describes how innovation at the front-line in healthcare delivery can be achieved by the very teams of people that deliver the care. It describes an example of how reflective practice, audit, innovation, research, development, and care process redesign were integrated to achieve a win-win-win outcome: a higher quality service for patients; a motivated and skilled clinical team; and improved performance with reduced costs to the NHS.

Who should read this book?
Anyone interested in improving healthcare services and designing them to meet the needs of patients should find this book useful. If your interest is in care pathway design and management of chronic conditions that span the primary-secondary care interface then it may be of particular interest. If you want to copy our model for shared management of leg ulcers then our experience should enable you to plan and implement a similar service that meets your local needs. This book is for everyone involved in delivering modern healthcare - nurses, doctors, managers, IT professionals, chief executives, ministers, the public, the press and most importantly the patients – our customers.

Why read this book?
Anyone directly involved in healthcare will testify that they are often too busy delivering the care to have the time to reflect, research and implement changes in the way they would like. Too busy to learn from the successes and the mistakes of others. Too busy to try different approaches and to evolve their own optimum solution. This book describes a range of ideas, methods, tricks and tips of how to avoid

the obstacles and pitfalls in the minefield of possibilities that is littered with the remains of previous failed attempts.

What this book does not do?
I do not wish to give the impression that the methods we used are perfect, complete or will work for everyone and for every problem. Every team has its own strengths and weaknesses; and every challenge has its own opportunities and threats. I do not advocate that all teams should develop their own methods and tools; in fact I would advise against it. However the general methodology and discipline that is needed to develop new tools is applicable to all design projects - especially healthcare process design.

Advanced information technology is revolutionising the way that healthcare is delivered. The more comfortable clinical and management teams are with the redesign process, the less stressful and more rewarding this change will be. Clinical governance is an organisation-wide concept; continuous quality improvement (CQI) is synonymous with process change; and all clinical and management teams will benefit from acquiring process redesign skills. I believe that these skills are complementary to audit and research and will in time become part of the training curriculum of all healthcare professionals: clinical and non-clinical.

How does this relate to other service improvement methods?
The principles and methods outlined in this book are encapsulated the term *Value Stream Design*. The *value* is what the customer wants and is prepared to pay for; the *stream* is the flow of work from a supplier to a customer; the *design* is the deliberate creative act of applying knowledge to create something that solves a problem.

Value stream improvement (VSI) is any method that seeks to improve both the quality and performance of a system by adopting a customer focussed perspective and working to identify and remove anything from the process that does not add value for the customer. VSI is not a new idea; it is the underlying principle for three schools of thought that over the last 50 years have transformed manufacturing industry; Lean Thinking; Six Sigma and Theory of Constraints. Unfortunately, these labels do not describe what they are and this ambiguity has resulted in some confusion and unproductive debate.

"Three Wins: service improvement using value stream design" is the same message told as a story; a true story. I hope you enjoy it.

Chapter 1. A new beginning

In the summer of 1999 a new department of Vascular Surgery was created at Good Hope Hospital to meet increasing demand and to satisfy national recommendations for a specialised vascular surgical service delivered by staff with appropriate training and experience.

Vascular disease affects the network of blood vessels; the arteries that carry blood from the heart to the body and the veins that carry blood back. Vascular disease affects a large proportion of the population, especially elderly patients, and is usually incurable and chronic and so places high demands on health care services. Close co-operation between primary and secondary care organisations is needed as complex surgical intervention is appropriate in some patients. Easy access to a vascular surgery outpatient service is required for patients to get specialist assessment, advice, and treatment; and many studies have shown that close collaboration between primary and secondary care delivers better outcomes in this complex group of patients and makes better use of the available human and physical resources.

Symptomatic arterial disease (e.g. hardened arteries) affects over 5% of people over the age of 65; and venous disease (e.g. varicose veins) affects more than 30% of the population. Around one person in 200 suffers with chronic leg ulcers; three quarters of which have a vascular cause. The modern management of chronic vascular disease requires detailed assessment by staff with specialist training and experience; access to sophisticated non-invasive imaging such as ultrasound; and the skills and experience of a consultant vascular surgeon. Such specialist outpatient services are only economic on the scale of one full time vascular surgeon per 150,000 population so a vascular surgery service requires the coordinated activity of a team of specialists that includes vascular nurses, vascular technologists, and vascular radiologists.

The majority of the demand is vascular disease affecting the legs and the typical problems that a patient will present with are leg or foot pain, and ulcers. The management of patients with these symptoms follows the same process as other patients; first the diagnosis (i.e. cause) is established and then the appropriate treatment is directed at the cause to relieve the symptoms and prevent progression of the condition. Most of this work can be done on an outpatient basis; admission to hospital is only required for the most complex and urgent cases and for patients that require specialist investigations or surgical treatment. Many of the conditions require shared management over a long period of time so accurate clinical records and good communication between the patient, primary care and secondary care is essential.

Establishing a diagnosis has three parts; taking a history by interviewing the patient; performing a clinical examination; and requesting special investigations such as blood tests, x-rays or scans. The most useful diagnostic test in many vascular patients is colour flow Doppler ultrasound (duplex), a sophisticated non-invasive imaging tool that requires specialised equipment and staff trained in its use.

A patient referred with a leg ulcer represents a complex problem because the management requires specialist clinical assessment, specialised diagnostic tests and specialist treatment with wound dressings and sometimes operations. The reason that leg ulcers are referred to vascular surgery clinics is because vascular disease is the underlying cause of over 70% of ulcers, and unless this cause is managed appropriately the ulcer will fail to heal. Leg ulcers affect around 0.5% of the UK population, cause untold pain and misery, and cost around £2000 per patient per year to treat; the main direct cost being the wound dressings and community nurse time. It has been estimated that leg ulcers alone consume around 2% of all NHS resources! The correct management of leg ulcers is well understood and under ideal circumstances and with expert shared care up to 70% of new ulcers will heal within 3 months. Other studies have consistently shown that the healing rate is only 20-25% when these patients are managed in isolation. Shared care delivers better quality.

The context

Leg ulcers are a good example of a chronic, non life-threatening disease that affects a large number of elderly people and which benefits from the collaborative care of community-based generalists and hospital-based specialists. The challenge in delivering a high quality leg ulcer service is not lack of knowledge of what we do but lack of application of that knowledge to how we do it. This problem is becoming increasingly common as the population ages because an increasing number of people have multiple, chronic conditions that span the traditional organisational boundaries of the NHS. Chronic disease management is the biggest challenge faced by 21st century medicine and to meet this challenge the balance of responsibility between primary and secondary care must be appropriate: primary care alone cannot achieve the desired outcomes; and secondary care alone cannot cope with the volume of work. The solution is to combine the strengths of the two services: primary care for day-to-day general assessment and treatment and secondary care for specialist assessment and treatment when required.

Good Hope Hospital (GHH) is a medium sized district general hospital situated on the north-east corner of Birmingham serving a mixed urban-rural population of around 450,000 and has for many years included vascular surgery as part of the department of General Surgery. The gradual development of the specialist vascular surgery service over the previous seven years had reached the stage where the time was right for the creation of a dedicated Vascular Surgery department. In 1999, the vascular surgery unit at Good Hope Hospital was created by increasing the number of consultant surgeons who had a special interest in vascular surgery to expand the existing team that included vascular technologists and specialist vascular nurses. Increasing demand for the specialist service had created major problems for the outpatient clinics, long waiting times for appointments and tests, disgruntled patients and increasing stress for the staff. It was not unusual for patients to wait four months for a new outpatient appointment, over six months for an outpatient duplex ultrasound examination and over a year for an operation. This story is typical of many specialist outpatient services that deal with chronic, complex conditions where the quality and performance of the service requires coordination of primary and secondary care teams and services that are designed to meet the specific needs of defined groups of patients. The effect of creating a dedicated vascular surgery unit was a predictable increase in the amount of work referred to an outpatient service that was already failing to cope. What was needed was not just more consultants - it needed a more radical and innovative solution to how we managed the whole patient pathway.

The challenge

The challenge was simple and one that is faced throughout the NHS: *We wanted to improve the quality of the service but we needed to do this within the constraints of existing resources.* Specifically we wanted to reduce the time that patients waited for clinic appointments, tests and operations; we wanted to reduce the number of times that the patients needed to visit hospital; we wanted to ensure that patients were cared for by staff who were competent to deal with their clinical problem; and we wanted to be able to communicate information and decisions quickly and effectively between everyone involved in the patient journey. At the same time we needed to increase the number of patients that were seen in clinic and we needed to do this using existing staff, equipment and facilities. In short we needed to achieve simultaneous "Wins" for the patient and for the NHS. The question was "How?"

The outcome

In July 2004 I had the honour of accepting, on behalf of the whole team, the first *NHS Innovation Award for Innovative Service Delivery* for the re-designed vascular surgery outpatient and leg ulcer telemedicine service. In April 2005 I was also delighted to accept the *HITEA Best Use of IT in the Health Service* for our innovative use of information technology in meeting this challenge. What is perhaps most surprising is that we achieved this nationally acknowledged success with none of the conventional service improvement machinery; there was no national directive, no business case, no project board, no special training; no service improvement experts and of course no money. There were just us.

A vascular surgery outpatient clinic is a specialised service and it is not the detail of what we did that is of general interest; it is how we achieved the success. This was the question that I was asked at the NHS Live Event in 2004. What surprised me at the time was that I couldn't give an answer; I couldn't describe how we had avoided failure. It would seem that the conventional methods used in clinical service improvement were not essential for success; and this raised a question in my mind "What are the essential requirements for success?" It is this question that I have been thinking and reading about since July 2004 and it is this question that I will attempt to answer.

The purpose of this book is to re-tell our story and to un-pick the principles and methods that we used. In doing this I have been forced to replay the twists and turns of the journey and to examine, with the benefit of hindsight, why we took the decisions we did. In retracing the steps I have gained a deeper insight into the process and into what, I believe, are the essential requirements for success. I have read some of the most widely quoted books on achieving successful change and have found a common set of principles echoed again, and again; though each framed in different contexts, using different methods and different language. My objective is to create a practical step-by-step guide of why and how to use the methods that worked for us and I hope this will help inspire others to gain the confidence to successfully overcome their own challenges. I believe anyone can do this. First you must believe that it is possible.

The goal

The first step is to define the goal. As Stephen Covey writes *"To begin you must have the end in mind"*. The goal of a healthcare system is to deliver the best possible service to as many patients as possible, and to deliver it when and where it is needed. In failing to achieve this goal we either deliver less than is required - a quality failure; or we deliver it too late; in the wrong place or at an unaffordable price - a performance failure. Our goal therefore has two clear objectives; we want to deliver the best quality and the best performance at the same time. However, there are three other possible combinations:

Lower Quality	+	Poorer Performance	= Lose-Lose
Higher Quality	+	Poorer Performance	= Win-Lose
Lower Quality	+	Better Performance	= Lose-Win
Higher Quality	**+**	**Better Performance**	**= Win-Win**

Our goal represents only one of the four outcomes; the other three represent a failure to achieve the one, the other or both of our objectives. To achieve the win-win outcome we must identify and eliminate the causes of these failures.

A quality failure is a problem with *what* we do and a performance failure is a problem with *how* we do it.

To achieve the win-win goal we must adopt a philosophy of eliminating errors and to do this we must find their root causes.

Three Wins

As a starting point it is helpful to list the eight possible *causal* associations between quality and performance:

Lower Quality ▶	Poorer Performance	Lose ▶ Lose
Poorer Performance ▶	Lower Quality	Lose ▶ Lose
Lower Quality ▶	Better Performance	Lose ▶ Win
Better Performance ▶	Lower Quality	Win ▶ Lose
Poorer Performance ▶	Higher Quality	Lose ▶ Win
Higher Quality ▶	Poorer Performance	Win ▶ Lose
Higher Quality ▶	**Better Performance**	**Win ▶ Win**
Better Performance ▶	**Higher Quality**	**Win ▶ Win**

Combining these into pairs there is one downward spiral of lose-lose, four stable compromises with mixtures of win and lose, and one upward cycle of win-win. I can think of actual situations that fit all of these categories; the question is why would you choose to lose?

> In our case we needed to increase the quality and performance of the vascular surgery outpatient clinic and that meant reducing the many opportunities for wasting patient's time and for errors arising from lack of information or poor clinical decisions. In a conventional clinic a new patient with a leg ulcer would visit the hospital three times before the diagnosis was established and definitive treatment could start - first for a specialist assessment, second for special tests, and third for a review with the test result. This process created two delays that added nothing to the patient care; multiple visits that cost the patient and their carers extra time, effort and stress; and added administration costs for the hospital. It seemed clear that we should start by looking closely at what we were doing and identify the root causes of the problems. We appeared to have an unhealthy process and we needed to diagnose the cause before we could decide the correct treatment!

It is often assumed that quality and performance are always opposite ends of a see-saw; what you gain on one you lose on the other. This assumption does not stand up to critical examination; if you can lose on both then you can win on both. This is not a zero sum game - you can have more than one winner.

Believing that a win-win outcome is possible is the most important step to achieving one.

The action

Even when your goal is clear and you believe that the goal can be reached the challenge of how to solve the problem may still appear insurmountable. After all, you managed to get yourself into this mess so what hope do you have that you can get yourself out of it? Where do you start? All the authors I have read agree on a fundamental principle in bringing about successful change - Action!

The guaranteed path to failure is to do nothing.

The only chance of success is to do something.

If you don't know the solution to your problem you have to ask two questions. "What is the problem?" and "What is the cause?" Whenever you are stuck or unsure the most useful thing you can do is to gather some information; ask questions, lots of questions, difficult questions.

I asked some questions: I asked "Why do we make our patients to visit hospital three times?"; "Why can't we do everything that the patient needed in one visit?"; "Would a One Stop Clinic improve the quality of care in terms of waiting times and reduce administrative costs?"; "Why aren't we offering One Stop Clinics now?"; "Why can't we just do it - it won't cost anything extra because we are already doing the same work now - just less efficiently?" I asked lots of questions and I got lots of answers like "Because we have never done it that way."; "Because we won't be allowed to."; "Because we haven't got time to organise it."; "Because it's not my problem."; "Because it's not my fault."; "Because it's not my job."; "Because it won't work.". I also got a few answers along the lines of "I can't see why that should be a problem."; "It would actually make my job easier."; "Why do we need permission?", and most importantly "Good idea, let's give it a go."

The asking

Other than making you feel better because you are actually doing something there are many other good reasons to start asking questions:

- It provides a clear signal that you want to learn and change.
- It provides a focus for action for every member of the team.
- It provides explicit evidence of where the problems lie.
- It provides a baseline from which improvements can be measured.

There are many different types of data that can be collected and virtually all of these will be useful at some stage so it is worth adopting this good habit early. Most data can be collected simply and quickly on paper *by the people involved* - patient views, staff views, activity, case mix, timings, pathways, literature searches, summaries of discussions with experts, etc. There is no need at this stage to conduct a detailed analysis of the data - all you need to do is agree what data to start collecting, how to collect it, who will do it and for how long. Do not assume that someone else should, could or would do this for you.

One of the first things we did was to start collecting data and I cannot stress too much how important this basic first step turned out to be for the success of the whole project. The questions I asked were simple, objective and the answers were easy to collect as part of the clinical process:
- When was the patient referred?
- What problem were they referred with?
- Who saw the patient in clinic?
- What investigations were requested?
- Where were the investigations done?
- When were the investigations done?
- When was the result available?
- What was the diagnosis?
- What treatment was offered?
- etc.

All these questions have two things in common: first they are all about the patient and secondly they are all "who, what, where and when" questions. The purpose of asking them was to record only what was

> happening not why or how. This involved some extra effort because this data is not normally collected routinely or is not easily accessible and it needed a convenient way of storing but it became clear very quickly that the benefits more than justified the effort. It does not matter if you collect the data on paper or electronically, but you should be selective. If you cannot answer the question "Why am I collecting this data and what benefit will it be to the patient?" then don't include it. To collect data for the sake of it in the hope that day it will be useful is a waste of time and we didn't have time to waste. That is not to say that you always collect the same data - circumstances change and new questions arise that need to be answered - so the questions you ask may need to change.

The listening

All the authors I have read also agree on another fundamental principle of bringing about successful change; *Listen First*. Change implies learning, so by asking questions and collecting data we are starting to apply this principle. The quickest and easiest way to learn is by asking questions and listening to the replies. The slowest and hardest way to learn is by making mistakes; and improving quality means making fewer mistakes so learning this way is not the logical path to improving quality of care.

Listening has two stages: hearing (collecting the data) and understanding (interpreting the data). The second stage is what many authors call *active listening* because it is only at this stage that learning can happen. It is important to appreciate that the interpretation of the same data will vary from one person to another. This is normal and expected because no two people have the same knowledge, experiences or make the same assumptions. So whenever sharing your opinion we should always expect some lack of agreement - and this is one reason why it is important to share our opinion *and* the evidence on which it is based; so that others can form their own opinion based on their own unique perspective. An opinion is just that, and when opinions differ there is no value in arguing which is "correct" – start with the facts and if there are none then start asking questions.

All authors also agree on a further principle of bringing about successful change; *Challenge Your Assumptions*. By active listening we test our own assumptions by focussing on the areas of disagreement. The only value of different opinions is to highlight the fact that there are conflicting assumptions. After reviewing the facts we may conclude that our assumptions are valid in which case we don't change our opinion. We may however conclude that one or more of our assumptions are invalid in which case when we change these assumptions we will have learned something and we will have changed. We may now find ourselves in agreement; our common ground has enlarged and we have progressed towards a win-win solution. Only when we believe that all our assumptions are correct, and we have evidence to support our belief, can we speak. Trevor Gay sums this principle of active listening in his phrase *"Listen but don't listen"* and Steven Covey in *"Seek first to understand before being understood".*

Often when we listen and challenge our own assumptions then many of the obstacles that we thought were in our path just disappear. Obstacles are often just mirages created by our own invalid assumptions. Whenever we find ourselves thinking "I can't do that" we are seeing an obstacle between us and our win-win goal, so we must ask ourselves "Why?" and keep asking "Why" until we have made explicit the assumptions that created the mental barrier. Sometimes this block is real; often it is an illusion. This simple method of asking "Why?" repeatedly is a powerful tool for uncovering unconscious assumptions. Ask yourself the "Why?" questions and listen to your own replies. Feeling stuck or helpless is just a state of mind and taking action by asking yourself "Why?" is a surprisingly easy way to make progress. Be mindful however that we find it uncomfortable to make our assumptions and motives explicit in public; in open debate it is gentler to start with "What, where, when, who and how?"

Our stated goal is to improve the quality and performance of the clinical service and the same principles apply to how we achieve this goal; we have to *Walk-the-Talk* and that means

1. Start with the end. Define your goal.
2. Do something. Ask questions.
3. Listen first. Challenge your assumptions.
4. Learn. Improvement means change.

Of course these concepts have been around for thousands of years, but their durability suggests there is wisdom we can gain from them.

The Tale of Three Hospitals

In response to externally imposed quality and performance targets, three similar hospitals decide to initiate service improvement projects in the two worst performing departments.

The first hospital decided to focus on the quality of the service by identifying and correcting the human errors through a process of no-blame reporting, sensitive investigation and focussed corrective action. The reduction in human errors in one department led directly to reduced work in correcting mistakes and an improvement in morale of the staff; both of which led to improved performance. However, in another department the focus on quality was interpreted as a "witch hunt" and led to a reduction in the morale of the staff, an increase in the time spent checking for possible mistakes and a reduction in performance.

The second hospital decided to focus on the performance of the service by identifying where long waits occurred, mapping the processes, rigorous investigation of the root causes and focussed corrective action. In one department the elimination of process bottlenecks reduced the long waits and cancellations, reduced frustration and stress, and led to an improvement in staff morale. In a different department the pressure to increase in performance was interpreted as "dead horse flogging" and was followed by an increase in errors, a reduction in the morale of the staff and eventually a fall in both the quality and performance of the service.

The third hospital considered that quality and performance were closely interdependent and that active management of both must occur simultaneously. The purpose and nature of the work in the two departments was analysed and both human and process errors were identified and corrected. One department responded well to an initial focus on quality and subsequently to improving the process bottlenecks that were then uncovered; the other department responded better to an initial focus on performance and subsequently to maintaining the quality of the service. In both cases the solution was matched with the root cause be it a quality or performance failure and by this balanced approach quality, staff morale and performance improved in both departments.

The purpose of this tale is to illustrate the four change principles being applied; the first having a clear goal - quality and performance improvement; the second the importance of action - the service improvement project; the third the need to listen first - identify if there is a quality or performance problem at the root; and the fourth the need to learn in order to change. The story also illustrates that blindly applying the four principles does not guarantee success. What is missing? Look at the first principle again - we are agreed *why* change is needed and have stated *what* the goal is, but we have not considered *how* a solution will be implemented. William Bridges describes the final stage of implementing a change as the *transition* and claims that much of the resistance to change derives from past experience of badly managed transitions. A transition has three stages: the *letting go,* the *neutral zone* and the *new beginning*. One of the critical parts of managing the transition is to allow those affected by the change to let go of old habits; and to do this we need to *sell the problem* rather than the solution. In other words, those who are affected by the change need to accept ownership of the problem before they will see the opportunity that any solution might offer.

> One of the first things that we did as a team was to take ownership of the problem, the need to find ways to deliver a better quality service to our patients. This appears to have happened before I joined the team, probably because the "letting go" had already happened when the new Vascular Surgery department was created. We were already in the neutral zone between the old ways of working and the new ways that had yet to establish themselves. William Bridges describes the neutral zone as both a time of threat and also a time of opportunity because it is a chance to consider new solutions to old problems; an opportunity to be innovative. We established a monthly team meeting and at one of these early meetings we used a team development technique called a SWOT (strengths-weaknesses-opportunities-threats) analysis as a way to encourage the whole team to talk about their hopes and aspirations; to be open about where the anticipated problems lay; to agree on where the priorities were; and to agree who was going to do what. As part of this process I accepted ownership of the problems in the outpatient clinic and the leg ulcer service simply because I had an interest in the cause and treatment of leg ulcers. To me it seemed sensible to start with the outpatient clinic because this is where the patient journey started for us.

The team

A team is just a collection of people who work together to achieve a common goal. Experience teaches us that some teams are more successful than others and there seems to be no guaranteed formula for developing and leading a successful team. The library shelves are crammed with books that discuss leadership skills and despite this there does not seem to be a foolproof way to identify successful leaders; except in retrospect from their track record of success. A team may be formed deliberately to meet a specific goal but more commonly a team evolves over time. Members join and leave for many reasons; the composition of the team changes; the challenges the team faces will change and the "leadership" role will change.

A successful team appears to be more that the sum of its parts; demonstrating a form of synergy where the differences between the members appear to be complementary and constructive.

An unsuccessful team appears to be less that the sum of its parts; demonstrating dysfunctional behaviour that is both antagonistic and destructive.

Over 20 years ago Meredith Belbin observed the performance of management teams that were created deliberately during training courses and found, to his surprise, that the most successful teams did not automatically result from combining the brightest and best of the participants. Based on individual personality profiles he found that successful teams included a mixture of members with personality traits that suited them to specific roles. In other words each individual had a natural set of strengths for particular roles, and by combining individuals and assigning appropriate roles according to their strengths the team as a whole became stronger and more than the sum of its parts - as Stephen Covey puts it - the team had *synergised*.

Conversely it is possible to achieve the opposite outcome; when all the members of a team share the same strengths or there is a mismatch between the individual strengths and their assigned roles. In this situation then either competition for roles will occur or members will fail to achieve their potential because their strengths become weaknesses in this context; the team becomes "unbalanced" and dysfunctional and less than the sum of its parts!

The Belbin Team Roles

- Plant
- Resource Investigator
- Monitor-Evaluator
- Completer-Finisher
- Team Worker
- Implementer (previously Company Worker)
- Shaper
- Coordinator (previously Chairman)

The insight provided by Belbin's work is particularly relevant to team performance in periods of change because different roles are required for the different stages of the transition process.

Successful teams were also those that were better able to deal with change and Belbin showed that this requires a definable set of roles linked to personality profiles; in other words successful change teams can be designed to some degree. However, mapping the personality profile of each individual in order to "design" a team or assign an appropriate role is similar to asking them to state their assumptions and motives in public; it is a personal and sensitive area which most people find intimidating. By applying the third principle of change - *Listen First* - the whole team can be encouraged to learn the principles of team roles and *ask themselves* the necessary questions to gain a deeper insight into their own strengths and weaknesses and then identify their most appropriate role in the change process; the role that plays to their strengths rather than their weaknesses.

Experience suggests that this concept of team roles is a bit too rigid because it implies that individuals will only play their specific roles. In reality, an individual is more effective if they can adopt a range of roles. They will have a preferred role that they find easier and more natural, and through practice they will develop greater flexibility, capability and an ability to perform well in almost any situation. This type of deliberate role shifting is well exemplified by Edward De Bono's *Six Thinking Hats;* a conflict-defusing technique that is easy to learn, and that speeds up decision making for an individual and a group.

> Some years ago, out of curiosity, I had my personality profile mapped and I "scored" highest on two dimensions - "Risk Avoidance" and "Innovation". Apparently this is an unusual combination as innovators are also usually risk-takers. The insight was more enlightening than threatening as it explained why I naturally avoid certain situations and find certain jobs irksome and unrewarding. In the challenge of improving the Vascular Surgery outpatient service it was these strengths that proved to be crucial - with my Innovator hat on I would naturally challenge the status quo and consider novel solutions, and with my Risk Avoider hat on I needed to have evidence that what was proposed would work before implementing it. While writing this book I used Belbin's Self Perception Inventory (SPI) to assess my strengths and weaknesses in the team roles and I came out stronger on the Shaper, Plant, Resource Investigator and Completer-Finisher roles which meant that someone else must have been compensating for my weaknesses in the Team Worker, Implementer, Coordinator and Monitor Evaluator roles. I have not asked the whole team to complete the SPI but I suspect that Sue Hayes, our research nurse, scores highly in those areas. Sue is the other half of the core team that steered the development, testing and implementation of the Leg Ulcer Telemedicine System. Sue's warmth, patience, tact and diplomacy, organisational and teaching skills are the perfect antidote to my impatience and intolerance that show through when progress is not fast enough or people do not demonstrate total commitment to the cause! Without someone like Sue on the team the project could not have succeeded.

The learning

Change involves learning and learning takes place before, during and after the change - at all stages of the transition. Just as the process of change can be painful so the process of learning can be uncomfortable. Change and learning are two views of the same thing. One widely used model of learning is the Conscious-Competence model where the path is described in four stages that are the combination of two factors; our insight and our knowledge. We all start with neither; we are not aware of what we don't know; we are not conscious of our lack of competence. This is the state of *blissful ignorance*. The next stage is entered when we become aware of our lack of knowledge, we become conscious of our lack of competence; either by making an error or by being asked a question that we can't answer. This is the stage of *painful awareness* and the transition from *blissful ignorance* is generally an unrewarding experience because no one likes to have their lack of competence demonstrated, particularly in public. The third stage is entered when we willingly accept new knowledge and learn; we are conscious that we have become competent and this is usually a positive experience but requires effort to achieve and maintain. This is the state of *know how* and only in this state can we teach others. The final stage is only entered after a period of practise through repeated use of the knowledge to the point where we no longer have to consciously think to use that skill; it has become *second nature*.

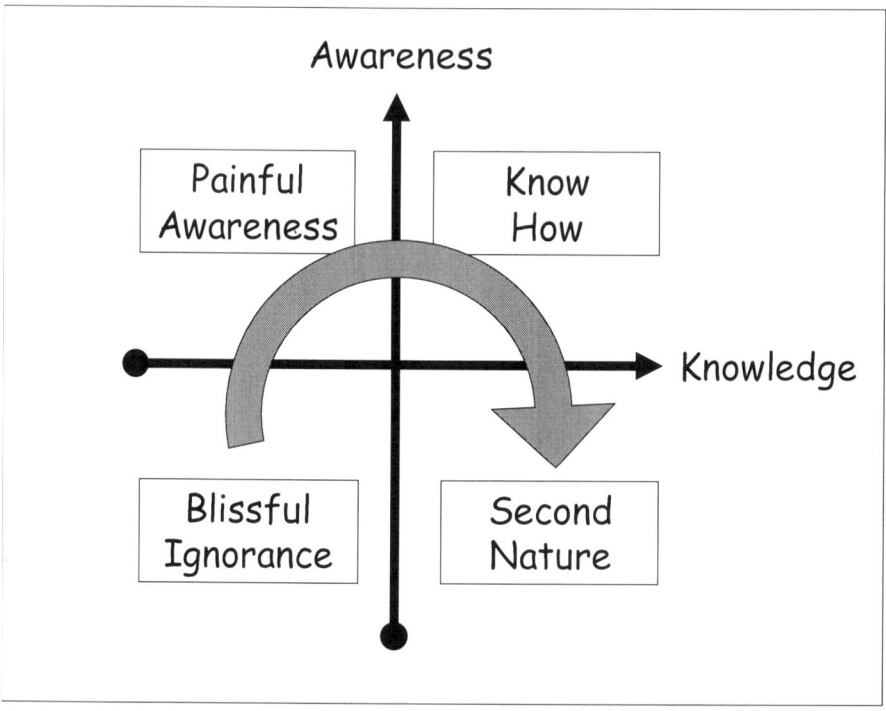

The Conscious-Competence model of learning.

Three Wins

> Some years ago I was given some juggling balls for Christmas. I was blissfully ignorant of how difficult juggling is and I assumed it would be an easy trick to learn. I was soon painfully aware of the truth; it is not easy – well not for me. I tried, got frustrated, gave up a few times, but determination got the better of me and eventually, after lots of practice, I succeeded. For me completing the path from painful awareness to know how brought a deep sense of achievement. For a while I still had to concentrate hard to avoid dropping the balls but with a lot more practice I can now juggle three balls without thinking about it; it has become second nature. My experience exactly mirrors the learning cycle; I knew that juggling was possible, I was motivated enough to put in the practice because I wanted to be able to do it, and the personal sense of achievement was the reward for the effort. This experience taught me a useful lesson; the rewards come from meeting challenges that are tough but achievable; so when building your own motivation start with problems that will stretch you but you know can be solved.
>
> In fact I now use the juggling exercise to demonstrate that very often the only thing that stops us from doing something new is our incorrect assumption that we will fail. Most people can learn to do a two ball cascade in just a few minutes with the help of a teacher. First the basic elements of throwing and catching are checked using one ball; then the new steps are demonstrated and "programmed" by simple conscious repetition with one ball; and then with two balls. In my experience most people do not believe they will do it and are very surprised when they discover that they can. It is like magic; and the demonstration is as powerful for those watching as those participating. If we can learn to drop one progress limiting belief then we can learn to drop others.

The three stages of a transition represent the changes from one state to the other on the learning pathway. We have all experienced the pain of change and it is particularly memorable when we are left in the neutral zone; the state of painful awareness; where we cannot or do not learn how to achieve the new competence. It is poorly managed transitions and frustrated learning that leave people in the painful awareness stage of the transition and creates the emotional resistance to change. To be forced to change and then offered no help in crossing the neutral zone is a sign of poor leadership. For successful change the whole team must *Walk-the-Talk* and that includes the leaders!

If the leaders are not challenging their own "comfort zones", asking questions, learning and changing then it is difficult for them to convince anyone else to do the same. This is the great risk of change – a growing credibility gap between the "tellers" and the "doers". If the gap is allowed to widen too far then the stage is set for revolution if the doers become impatient; or regression if the doers lose heart. The way forward is evolution – the doers and tellers challenging each other and learning from each other.

Chapter 2. Win-win-win

We want to improve the quality of our healthcare services and we also want to improve the performance of these services but the issue seems to lie somewhere in the link between quality and performance. The problem seems to be in how we work together as a team rather than what we do as individuals; we seem to be doing the right things but we are just not doing them the right way.

> In our case we were already painfully aware of the need for change - we were in the neutral zone and we were searching for a way forward. There was no way back. We were agreed on the goal; we had taken ownership of the problem; we had identified our strengths and weaknesses; and we had started to ask questions and to collect data. The next stage was to identify the root causes of the problems and to generate some options to fix them. It is often said that 80% of the solution is in stating the problem and in our case the problem was most apparent with patients suffering from leg ulcers. They were getting the worst deal; long delays for appointments and tests, multiple visits to the clinic, poor coordination of their treatment and inevitably poor outcomes. Looking at the problem from the perspective of the patient the problem was clear - poor communication. There was poor communication within the hospital because diagnostic tests required a separate visit to a different department. There was poor communication from visit to visit because it was not clear from the hospital notes what progress had been made since the last visit and if the treatment was working or not. There was poor communication between the hospital and community teams because it was common for patients to be referred only when the community team had run out of ideas and also common for treatment to be changed by either team without a clear rationale. We were all working to national guidelines for managing leg ulcers and still delivering an inadequate service. What was needed was a shift to a patient-centred solution with a strong emphasis on improving communication and reducing delays for patients. The benefits for the patient would be a shorter wait for an appointment, better coordination of the effort of all members of the health care team, better decisions based on complete and accurate information, reduced number of visits to hospital, and hopefully faster healing and an improvement in their quality of life. We needed to fix the process. We needed to fix how we were delivering the service not what we were delivering. This was the goal we thought.

In healthcare there must be a process that links people together to create a team that can deliver a high quality service. However, the process also includes the same people that feel the pain of change; and the same people who accept ownership of the problems; and the same people who accept the challenge of improving the service. Any solution that delivers a win for quality and a win for performance must also deliver a win for the people involved. There is a problem with the win-win principle; it is not enough.

We need Three Wins!

The processes are passive, it is the people who are the active agents; and without both working well together there will be no third win. The goal is a win-win-win solution. The challenge is to design, test and implement processes that work for patients and people; and to ensure that the people have the necessary skills and resources to do what the patient and process requires. We need to do the right things and we need to do them the right way; and to achieve that the people involved must feel motivated.

The motivation

The subject of motivation has attracted a lot of discussion in the literature and I will only mention two widely quoted theories that provide an insight into the roots of individual motivation; the Maslow Hierarchy of Needs and the Herzberg Theory of Motivation. The essence of both is that an individual's motivation is determined by a number of factors which can be divided into two groups; the *essential* and the *desirable*. Essential factors do not lead to motivation themselves but if not present will prevent motivation and lead to dissatisfaction. Maslow described these as the *physiological, security* and *belonging needs* and Herzberg called them *hygiene factors*. Surprisingly these were found to include things such as the working environment, salary, status and working relationships. The desirable factors are responsible for genuine motivation and were found to include things such as having an interesting and challenging role, responsibility, recognition of achievement, and opportunities for personal growth and advancement. These two classes of motivation factors can also be thought of as *survival* and *growth* factors. Individuals need to feel safe and secure before they can explore their own potential and grow to achieve it.

> The sources of motivation for the Vascular Surgery outpatient team were clearly different for each individual and it was interesting to observe that some members demonstrated more motivation to change than others. For me it was the challenge of the problem and the feeling of achievement that comes from delivering a successful solution to a difficult problem. The other members that showed the greatest motivation were the specialist vascular nurses, and their motivation seemed to derive less from the challenge and more from the potential for personal growth and greater responsibility. A considerable number of people in all parts of the organisation appeared to show little or no motivation. Why was everyone not equally committed to improving the situation? What threat did changing the way we worked pose? With the benefit of hindsight, this variation in motivation is predictable, understandable and perfectly reasonable; at the time it was the source of considerable frustration! What I experienced from the majority was passive support for our proposals; everyone thought it was a good idea but no-one offered to help do it; and that included managers and colleagues. I now know it could have been much worse; I could have experienced passive resistance or even active resistance! Interestingly the participants that showed least motivation were the ones that viewed the project only as a way to reduce their dissatisfaction with the current situation rather than to achieve personal growth. I learned several lessons from this. When defining the goals it is important to consider what each individual will gain from the outcome; if it is just satisfying a survival factor then expect lack of resistance but little assistance; if it is satisfying a growth factor then expect offers of active support; if it is interpreted as a threat to either then expect trouble. The converse is also true; if you need assistance then you need to build motivation and this means the proposal must offer opportunities to increase security and satisfy individual needs for growth. One of the purposes of a SWOT analysis is to enable individuals to explore their own motivation and when done well you won't need to ask for help - it will ask for you.

The objectives of a win-win-win outcome are now explicit:

> **Win = Improved Motivation**
> **Win = Improved Quality**
> **Win = Improved Performance**

Three Wins

The first stage of the win-win-win journey is to establish a clear purpose, a clear vision and to get everyone started on the journey to new knowledge with the more motivated people taking action and leading the way and the less motivated observing and following later. All we need now is a plan.

> A strength that proved particularly useful at this stage was my knowledge of information technology. While at university in the early 1980's, just at the time when the personal computer revolution wqs happening, I saw a golden opportunity to get formal training in computer science; an investment that has paid dividends. One of the problem-solving techniques taught is called *Divide-and-Conquer*; the principle that complex problems can be broken up into smaller, simpler, solvable parts, and the components then combined to solve the original problem. This principle is widely applied in many fields and appears in many guises; one of which is the project lifecycle that in computer software development is described as a Seven Stage process. The first three stages involve breaking the problem up into parts, the fourth stage is the detailed work of solving the parts, and the last three stages are the testing and integration of the parts to build the solution to the original problem. This discipline is critical to successful development of complex computer software and was therefore a skill that I had learned and practised and so to me was second nature.

The plan

A plan is a set of defined tasks linked into a logical order; the output of the divide-and-conquer principle applied to the problem. A plan is just one possible path between the "now" and the "future". However, it is likely that there are a number of possible paths and the challenge is to decide which paths are feasible and then decide which of the feasible paths to take. Change means learning; learning means being curious, and being curious means challenging assumptions. The more innovative we are in our questions then the more options we will generate and the more paths there will be to consider. Applying the "*Start with the end in mind*" principle it makes sense to find the *easiest* path - the one that has the greatest chance of success for the least effort. If there are obvious, simple, quick, beneficial changes that can be done immediately then apply the *Action* principle - Just Do It.

Our objective was to improve communication at all parts of the patient pathway and to find ways to reduce the number of separate steps that a patient had to complete and thereby eliminate delays from the process. One innovative option was to change the conventional multi-visit clinic (MVC) to a patient-centred One Stop Clinic (OSC) because this would help with communication and eliminate about 12 weeks of delay for the patient without increasing the workload for the staff.

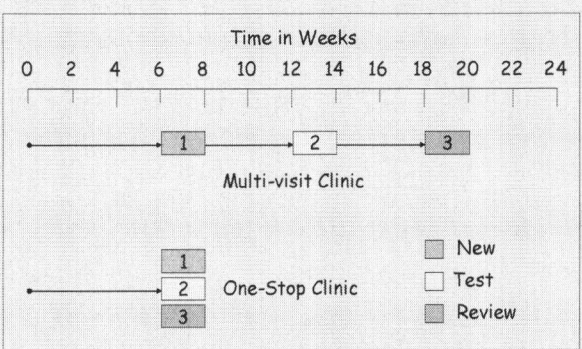

The idea of a One Stop Clinic seemed simple and obvious but was met with some resistance because it would require bringing the necessary staff and test equipment (the duplex ultrasound machine) from the radiology department to the outpatient clinic and performing the scans in a more ad hoc way compared with the conventional radiology booking process. In other words it was a significant change to the way that the radiologists and vascular technologists worked. But there was a carrot: patients with leg ulcers are inconvenient in the radiology department because when the ulcer dressings are removed to do the tests, the staff in radiology do not have the skills to redress them correctly and have to call for specialist vascular nurse assistance. The One Stop Clinic was also a solution to this long-standing niggle for radiology and the mutual advantages for patients, clinic staff and radiology were so obvious that they overcame the resistance to change. This simple but fundamental change was achieved quickly and easily and had a massive impact - the time from referral to definitive diagnosis was reduced from 18 weeks to just 6 weeks and two visits to hospital for the patients were avoided. The lesson was clear: even small changes can deliver big win-win-win outcomes; the trick is to consider the problem from all perspectives and identify the simple changes that have potential benefits for everyone involved.

A question that I am often asked when I present this story is "Haven't you shot yourself in the foot by doing the same work in fewer outpatient visits because you'll generate less income?"

The administration and supporting bureaucracy tends to align itself around the current situation; and in healthcare the income generated is determined by the number of visits; new appointments attracting a greater fee for service than a review visit. Of course, this arrangement tends to perpetuate the behaviour of multiple visits which, as we have seen, are not in the interests of the patients, staff or the organisation.

So by improving the quality of care and reducing delays by implementing a One Stop Clinic we created a solution that was now out of step with the bureaucracy – and bureaucracy does not change quickly. What we did not want was for it to force a regression back to the old way of working. Our solution was to book the patient separate clinic New and Review appointments; just on the same day at the same place and at different times. This way the activity was recorded accurately and a potential conflict was defused.

One day the bureaucracy will catch up with reality; that is a challenge for the future. I believe it is called Activity Based Costing (ABC) but I'm not an accountant.

Chapter 3. Innovation

Innovation means *new*. Anything that is new to you is innovation even if it is not new to someone else. Something that is new for everybody is called invention and a truly novel invention is a rare and precious thing. The difference between what is new and what is novel highlights one important principle of good innovation: *Don't Re-invent the Wheel*. When we have a challenge and need to be innovative we should first assume that the solution already exists and it is just a matter of finding it. To do this we must have a clear idea of what we want, be prepared to do a bit of searching; be prepared to challenge our assumptions; be prepared to ask difficult questions; and be prepared to listen to the answers. The place to start this search is us. It is surprising how often we know the answers to our own questions if only we bother to ask. We can use the *Repeated-Why* method to uncover our unconscious assumptions and motives.

Some people have a natural ability to think laterally and to generate new ideas; Meredith Belbin identified this important role in successful teams and called it the *Plant role* (the term derives from the fact that he 'planted' people with this ability into teams to observe the effect on team performance). If you don't have a Plant then the easiest place to find solutions is by looking outside the team; ask around, find people who have a reputation in the subject, ask them, ask them to point you to other sources of information. Be sure to record your findings at each step - do not try to remember what you hear - some of these ideas will be helpful and will become part of your new knowledge and others will not; just record what you hear and reflect on it later or just pass it on. Some people are naturally attracted to this innovation foraging behaviour and Meredith Belbin called this the *Resource Investigator* role of successful teams.

> One essential attribute of an innovative team is curiosity; the desire to seek out the unfamiliar and to learn from it. Just being receptive to different ideas builds a mental store of things that "might come in useful one day". Curiosity has another advantage; what appears to be a new idea on closer inspection can turn out to be something familiar in disguise. This simplifies things a lot. We can be innovative without being inventive; we can be curious; and we never know when an idea we picked up one day might come in handy!

The brainstorm

There is a technique called *brainstorming* that a team can use to make the best use of their own creative powers. Everyone is able to generate ideas but are often inhibited from sharing them by lack of confidence and fear of ridicule if someone else thinks their idea no good. The purpose of brainstorming is to allow all ideas to come to the surface and very often just expressing an idea acts as a stimulus for other, often better, ones. The method involves assembling the group, explaining the "rules of the game" and then asking one person to act as facilitator. The first task is to agree the goal of the session and the measures by which ideas will be judged. The rule for the first half of the session is "do not judge ideas until all ideas have been suggested" because what may appear as a poor idea initially may ultimately turn out to be a good one when all views are taken into account. Getting a brainstorming session running sometimes needs some challenging questions or an injection of ideas and this is where the *Plant* and *Resource Investigator* roles can contribute. Brainstorming is an exercise in challenging assumptions and must therefore be done in a supportive and non-judgemental context. The rule for the second half of the session is that the whole group discuss and filter the ideas that have been generated and eliminate the ones that don't meet the objectives, leaving a set of ideas that do; the Options. You are not looking for the option that everyone agrees is the best; you are just excluding the ones that everyone agrees are the worst - an important distinction. When facilitated well, brainstorming is a powerful motivator because the best options are usually those that hit the most *growth motivation factor* buttons for everyone!

> We had already used a form of brainstorming, the SWOT technique, to define the purpose and vision of the team and to identify improved communication as our primary objective. The next step was to identify possible options to improve communication and to do this I applied a standard design method from my training in computer science; I started by defining the essential and desirable requirements – just like a person specification for a job. Starting with a blank piece of paper I just focussed on what we needed rather than how to meet those needs because experience has taught me that it is a mistake to jump to solutions before you have finished defining the problem. Even worse to start with a solution and manipulate the problem to fit. Our most difficult problem involved patients with leg ulcers; and the essential requirements were to be able to measure the response to treatment

and to communicate the plan of management to other members of the extended team even when the patient was not present. In addition our solution needed to be secure because it involved confidential patient information and it needed to be quick and intuitive to use by a wide range of healthcare staff. Finally we needed a solution that communicated actual patient data rather than just an interpretation of the data and clinical images were one way to do this. A number of possible options were immediately eliminated because they didn't meet these essential requirements; for example a patient-held paper record would not allow communication when the patient was not present; and e-mail was not secure enough and would not allow shared access to the patient information. The only option that met the essential requirements was a shared electronic patient record that included the ability to store clinical images; in other words what we needed was a telemedicine system. Having defined what was needed I then applied the *Don't Re-invent the Wheel* principle and conducted a survey of what was already available. Unfortunately this started to throw up some obstacles; first none of the available systems were designed for specialist nurses looking after leg ulcer patients; second there was no evidence that the available systems actually worked in practice; and finally they were too expensive to consider buying just to see if they worked or not. An alternative path was to find someone to develop a telemedicine system to our specification but this also threw up obstacles; first we didn't actually know what we needed in enough detail to write a specification; second there was no one available in the hospital to do this for us; and third we didn't have the resources to commission someone outside to do it. The last option was to do it ourselves, and although this might appear to be the least attractive option it turned out to be the only feasible one for several reasons; first it would allow us to learn what we needed in stages; second because it allowed us to evaluate progress at each stage; and finally because we actually had the skills to do this within the team. Luckily it was one of our strengths. The downside was that this option would take a lot longer to do but it was still our best option.

The map

Innovation is all about generating options; alternative paths to the future goal. These paths define the map of the terrain over which we must make our win-win-win journey. Some paths look very promising but turn out to be blind ends; some paths have obstacles in the way and we don't have what is needed to overcome them; and other paths may seem difficult at the start but turn out to be the only way to reach the goal. Of course it is possible that there are no clear paths from the start all the way to the goal; do not give up yet; remember the *Action* principle. There is always something you can do. You may not be able to see a complete path but you can work out how far you can get. Just set an intermediate goal that will stretch you but which you know you can reach and start working towards it. By the time you get there the path onwards is likely to have become clearer and some of the obstacles that you perceived at the start may now be manageable, may have disappeared and new paths may have appeared. It is better to achieve something easily and quickly that has a tangible benefit and which moves you along the path towards your goal than to give up because you can't see a complete solution. This is called the *low hanging fruit* principle – do what you can do. Easy wins are important because they demonstrate progress is possible, they demonstrate action gives results and they build experience, confidence and motivation. These early wins should be recognised, acknowledged and used to help overcome the obstacles in the path ahead. Each step in the right direction is a positive and valuable action - you may not be able to see all the way ahead but if you keep moving and keep your eyes and ears open for threats and new opportunities then you are much more likely to reach your destination than never leaving the safety of the status quo.

The technique of planning a series of incremental changes to improve a process is safer and more effective if you collect and analyse data to monitor the effects of the changes on the behaviour of a process.

The Plan-Do-Study-Act cycle (PDSA) was developed in the 1930's by Shewart and Deming in the USA and put to very good effect by manufacturing industry; notably in Japan during the 1950's. Only recently have these simple, tried-and-tested methods have been introduced into healthcare where they have been shown to work well provided the problem and solution are owned by the people who provide the care.

The PDSA cycle is simple action-oriented method; all it requires is that the objective is clearly stated and measures are agreed to test if the objective has been achieved (plan), that a change is agreed and then implemented (do), the effect of the change is measured (study) and any minor adjustments made based on the experience gained (act). The PDSA method is good when there is clear consensus on the changes to be tried, when changes are relatively easy and quick to implement, and when single changes will result in a measurable improvement.

> Our goal of improving communication by using a shared electronic patient record system was not going to be an easy win - but there were other things that we could do and did do. Changing the clinic from a Multi-Visit Clinic to a One Stop Clinic was relatively easy, offered obvious benefits for everyone, and cost nothing to implement. This was a piece of low hanging fruit and it turned out that we used the PDSA cycle to manage the change - although I had not heard of PDSA at the time. Remember that one of our first actions was to start collecting data; so by the time we implemented the One Stop Clinic we had enough baseline data to assess the effect of this change and we also had an established data collection method that would allow us to measure the improvement. The effect was dramatic - for patients with leg ulcers the time from initial referral to diagnosis was reduced from 26 weeks to just over 6 weeks and the number of clinic visits that the patients needed to make was reduced by 40%! However, the One Stop Clinic was not just for leg ulcer patients; equivalent benefits were demonstrated for other vascular surgery outpatients as well, such as those with arterial disease and varicose veins. This relatively simple change also increased the motivation of the team because we had achieved significant improvement with minimal effort; and had eliminated a lot of the stress that was being caused by the old process - the first step on the path to our win-win-win goals.

The luck

It should not be underestimated how much impact both good fortune and bad fortune can have in change projects. Vague threats can turn into real obstacles; unforeseen events can scupper the best laid plans; apparently solid opportunities can evaporate and new opportunities can unexpectedly materialise. The safest strategy is to keep moving, keep your eyes and ears open for opportunities and threats and to be prepared to change your plan quickly and decisively. An early SWOT or brainstorming session will identify your preferred option and a list of acceptable alternatives that may become useful as contingency plans. As Louis Pasteur once said *"Chance favours only the prepared mind"* and what might at first look like a show-stopper can turn out to be an opportunity in disguise. All change projects have set-backs: it is normal and is expected so everyone involved should be kept fully aware of this. In my experience people are more reassured by seeing unexpected problems identified and resolved quickly, effectively and openly than by seeing obvious and predictable issues ignored, missed or fudged.

> We were unexpectedly presented with an opportunity quite early in our project - the R&D Department at Good Hope Hospital circulated a letter inviting clinical teams to submit research proposals. Based on past experience I assumed that we would not get any financial support for our proposed changes so to seize this opportunity meant that we would need to re-design the project as a research study rather than a PDSA cycle. A research study requires a clear hypothesis, a robust and feasible method to test the hypothesis, ethical committee approval, dedicated resources to ensure the study is conducted to meet the necessary standards required, and ideally prior experience of running similar studies. It was clear that the research route was a sensible way to make a lot of progress in achieving our goal and in fact it was a better way to approach the problem because I could find no published evidence that telemedicine had been used successfully in the shared management of patients with leg ulcers. We had a hypothesis that needed testing and to test it we would only need to have a prototype telemedicine system and dedicated research staff to coordinate the study: both of these were achievable objectives if we had funding. I had already considered a research study as an option but had eliminated it because my experience of obtaining funding for research is a lot like playing the Lottery - you can spend a lot of time and money trying and still end up with nothing. Funding agencies quite reasonably favour researchers with a proven track record - but how do you get a track record without funding: a classic Catch-22. However, the letter from R&D improved the odds of getting a study funded so I prepared and submitted a research proposal for a small grant to allow us to recruit a part time research nurse for two years. The proposal was accepted and funded and allowed us to move forward along a path that I had originally assumed would be blocked.

Chapter 4. Investigation

Innovation should generate lots of ideas and options event though only one option can be implemented. The next challenge is to select the option that will be implemented. This can be approached in different ways; we can choose any of the options that will achieve the stated objectives but this means we need to have some way of knowing which will work and which will not; alternatively we can attempt to identify the *best* option and to do that we need to have an objective way of comparing the different options. There are of course other ways to select an option; we can use the *Who-Shouts-Loudest* method and fight about it until only one protagonist is left talking. Or we can just guess. Tossing a coin at least has the advantage of being easy, quick, cheap, fair and non-confrontational but has the disadvantage that it does not reliably identify the best option! Whatever method used we are trying to filter the suggested options to leave only one and when there are no previous successful examples to use as a guide then there is no guarantee that any of the options will work. In addition, even if a method has been shown to work elsewhere there is no guarantee that it will work for us. So to reduce the risk of failure we need to conduct some form of investigation to identify an option that has an acceptable chance of success. Of course if we are just repeating what many other people have done already then we can be more confident that it will work and we can learn from their experience and give our change project a better chance of success. Why try to re-invent the wheel and why make the same mistakes twice? The greatest threat to success at this stage is the *Not-Invented-Here* attitude; the assumption that if someone else has a solution then we should repeat the work and develop our own way of doing the same thing. Surely it is more sensible to build on the success and experience of others?

We had the advantage that we were a small team and did not have many options to consider; but we had the disadvantage that no one had successfully implemented a shared electronic record system for nurse-led management of leg ulcers so we had no one to learn from. Our project was going to be a step into the unknown and a formal research trial was a wholly appropriate method for our investigation and for the next stage of the journey to our win-win-win goal.

The filter

There are many ways of conducting an investigation to decide which options are feasible and which are better than others, but the method chosen must be appropriate for what we are trying to achieve; *we must use the right tool for the job.* In a situation where there really is only one option and this represents a relatively minor change a *Suck-It-And-See* method like PDSA is entirely appropriate. Trial-and-error methods might also include a prototype, a feasibility study, or a pilot study and are designed to collect information to assist the decision process or to plan a definitive investigation. The most complex and usually most expensive method is a controlled research trial which is a formal comparison of one or more options. A research trial is conducted in a rigorous, objective and statistically valid way that is designed to answer a specific question; "Does this work?" Whichever method is selected for the investigation it represents a separate sub-project and is therefore conducted according to the same principles. There are tried-and-tested methods for conducting trials that emphasise the need for discipline and attention to detail; this is not the time for hot-headed creativity and radical innovation. Consequently the skills required to conduct an investigation to a high standard are different from those required to generate lots of new ideas. Meredith Belbin observed this in successful teams; there was a role that showed critical appraisal skills, the *Monitor-Evaluator,* and a role that favoured attention to detail, the *Completer-Finisher*. He observed that the team members who are good at finding or generating new ideas are not always the best ones to investigate if these ideas are feasible or practical. Missing out the investigation stage runs the risk that new ideas are not tested enough before they are implemented - resulting in a predictable, wasteful and avoidable failure and a increase in the resistance to further change.

> We were now ready to start the journey towards our objective of using a shared electronic patient record to improve communication. However, before we could start the research trial we needed a working prototype and we needed to show that we could use this to accurately measure the response of a leg ulcer to treatment - i.e. the rate of healing. We had already used the divide-and-conquer principle to break up the bigger problem into smaller bits and here I was on familiar ground because developing computer software has a well proven process. Why re-invent the wheel?

The Seven-Stage design process is generic and can be applied to essentially any development project with virtually no modification except to change the terminology to view it from different perspectives. For example from top-to-bottom it can be viewed as "strategic-to-tactical" and left-to-right it can be viewed as "subjective-to-objective". The reason why it is so useful in computer software development is because it dramatically reduces the chance of error. A computer program only does what it was designed to do and if you don't specify what you need then you it will not do what you want. Computer programs don't learn, evolve and modify themselves like people do. If you don't specify exactly what you need then a software engineer can't help you because to write a computer program you have to break the actual problem down into a larger number of very simple instructions that a computer can understand.

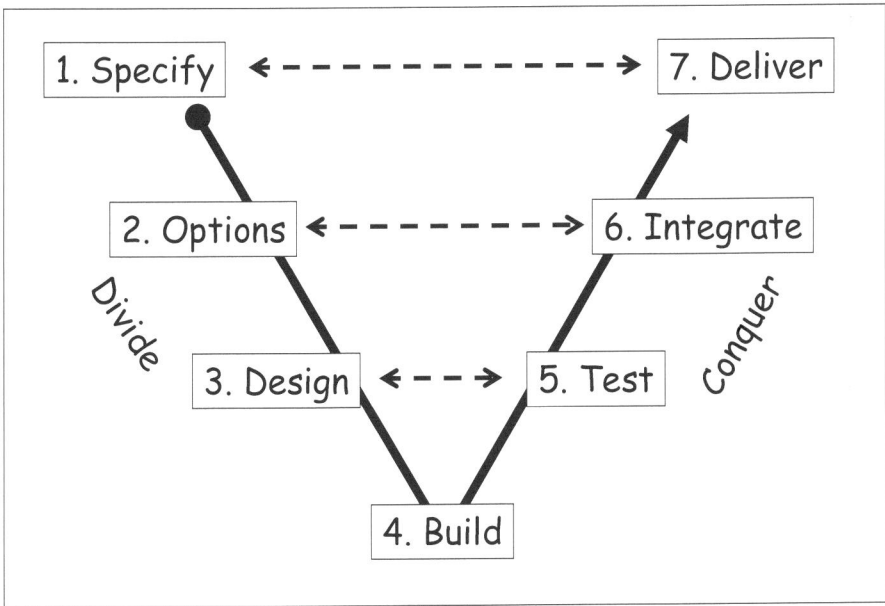

The Seven-Stage process of software design and development.

The requirements of our prototype telemedicine system were modest; we just needed to capture and store images of leg ulcers; use these to measure the size of the ulcer; show the change in size over time; and allow secure remote access to these records for anyone involved in the care of the patient. Using this clear description of what we needed and with modern software development tools this did not take too long

> to do and it gave us something tangible we could start evaluating. The recent availability of affordable digital cameras meant that it was now easy to capture and transfer images directly into a computer so the first task was to formally test the prototype ulcer measurement method. This step had another very important purpose; to get the whole team involved in the specification and evaluation of the electronic patient record system; and to give them emotional ownership of the solution as well as the problem. We did this by conducting a small study comparing the accuracy of wound measurement using conventional pen-and-paper methods with the prototype computerised system. The results showed that the methods were equally accurate but the computer method was much quicker. This was an important finding and we marked our achievement by publishing our findings. Publication has two important purposes; to disseminate new knowledge and to provide tangible evidence of personal achievement.

The diffusion

The first principle of change is *"Start with the end in mind"* and implies asking the question *"When we have finished this change project how will we disseminate what we have learned?"* It does not matter if the innovation worked or not, both outcomes represent useful new knowledge and should be disseminated to avoid re-inventing the wheel (wasted effort) or making the same mistake again (wasted effort). The second principle of change is *Action* and at the very start of a project this means being prepared to ask questions; the third principle is *Listen First* which means that someone else must be prepared to share their knowledge with us and answer our questions. If we were helped by others who were prepared to share their knowledge then we should be prepared to help others in turn and that means passing on what we have learned.

It is helpful at this stage to understand the process of how new ideas are generated, spread and taken up by others. The *Diffusion of Innovation* principle was described by Everett Rogers who studied a range of innovations in different contexts and observed how they spread. Rogers showed that individuals, teams and even organisations could be classified according to how receptive they were to new ideas on a scale that ranged from very receptive through ambivalent to very resistant. Rogers observed that there was a bell-shaped distribution of this receptiveness to innovation within a population and this led him to describe populations in terms of four groups that he called *Early Adopters (16%), Early Majority (34%), Late Majority (34%) and Laggards (16%).* He showed that most new ideas originated in a sub-group of the early adopters which he called the *Innovators (2%)* and these new ideas spread through the population in the order of receptiveness - from early adopters through the majority to the laggards.

Rogers' curve that shows the distribution of receptiveness to change within a population. Innovation diffuses from left to right.

Rogers' insight provides strong justification for deliberately disseminating new innovations and knowledge as widely as possible because within any organisation we expect to find the same distribution of *receptiveness to change*. If an idea is novel then the group that will be most receptive to the new knowledge will be the early adopters: There will be some of these within the parent organisation but most will be outside. For an innovation to become widely adopted it may diffuse faster outside the parent organisation than within and this observation provides the explanation for the oft quoted phrase *"You cannot be a prophet in your own land"*. Innovators and early adopters should not be put off by the fact that their ideas are not immediately taken up and disseminated throughout their whole organisation; according to the Rogers' Curve you would not expect it to! The most change-averse group, the laggards, can also be divided into two groups that we can label the *die-hards (14%)* and the *dinosaurs (2%)*. The die-hards are the last to change and take the most effort to convince; the dinosaurs will never change fast enough to keep up and like their namesakes have made their choice of destiny!

The root cause of this behaviour is individual motivation though it is incorrect to assume that individuals behave the same way for all new concepts. It is quite possible for an individual to be an innovator in one context and a laggard in another. The insight provided by Rogers' work is very useful when linked with the principles of motivation as it suggests that for change to happen quickly and effectively the barriers to *Diffusion of Innovation* need to be lowered; and this implies that we must build the motivation to want to change. The first step is to show change is needed, the second to show that change is possible; and the third is to broadcast this message to everyone. However, everyone will interpret the message in a way that is related to their perspective of the problem and I have heard it said that to communicated your message you need to *"Say it seven times in seven ways"*. In other words our message needs to be translated into different forms of language that different people will find easier to interpret and so that they have the best chance to see it as an opportunity for them; to enlist their support and motivate them to engage.

We had reached the critical stage where we needed to spread the word and promote change by "selling" both the problem and our proposed solution. We had a vision, we had a motivated team within the outpatient clinic, we had financial support for a research project and we had a prototype system that we could use to test the idea. We now needed to engage potential partners in primary care - and to do that meant finding the early adopters. We took the approach of talking directly to the people in primary care who stood to benefit most from the change - the district nurses. We explained the problem as we saw it and our proposed solution and how we thought it could benefit them and their patients. We offered to visit them and present the project to their teams and to listen to their comments and suggestions. As we already had funding for the project all we were asking for was their commitment to participate and the first teams that said "yes" we took as the self-proclaimed early adopters and we invited them to join the study. There were other important stakeholders in primary care, the GP's and PCT managers, but they had less to gain directly from the research study and so had less reason to be motivated to help. However, they were potential obstacles so it was important to approach them, explain the problem, explain what we and the district nurses wanted to do and explain that we didn't need their direct help but would be grateful for their support. In other words we didn't give them a good reason to say "No" - and we got their support. The only thing left to do was to demonstrate that we could successfully create a new communication link between a community nursing team and the specialist nurses at the hospital using a shared electronic patient record and the fact that this was now a research project actually made it a lot easier to get the necessary help and technical support.

The other essential requirement of the new system was that it should be secure. By applying the same principles I worked through a list of possible options and chose the simplest that would provide the appropriate level of security while still being robust and easy to use. While this work was being done we took advantage of another opportunity that came our way - to practice our research methodology by doing a commercially funded research project on leg ulcer treatment. This was not part of our original plan but it offered definite benefits - it allowed me to recruit a full time research nurse and to foster a more research-style culture within the department. It also attracted significant income and kudos for the R&D department - another unexpected though welcome outcome.

It would be wrong to give the impression that everything was plain sailing - we also suffered setbacks. The most serious at this stage was the departure of one of the most enthusiastic and dedicated members of the team, a member who had been highly influential in setting up the vascular surgery nursing service and was nationally acknowledged for her work. However, the inevitable changes that followed the establishment of the new vascular surgery department meant that everyone had to accept a lot of change and sometimes this creates the opportunity to leave to take on new challenges. William Bridges highlights that all new beginnings start with an ending. This loss was unfortunate but we had no choice and we had to keep moving forward.

The integration

A complete solution usually requires several components that must work seamlessly together. Some components are "off the shelf" and others need to be created from scratch and thoroughly tested before being incorporated. Eventually all the components need to be linked together or *integrated* and tested again as a whole to see if they are capable of achieving the stated objectives; this is called a feasibility study. There are still often several possible ways to solve the problem and the feasibility study can be used to compare these options and decide the single best option for the final stage of working towards the final solution. Just as with teams, the whole is greater than the sum of the parts, and often a successful solution is the result of the whole and not just one of the parts. It can be difficult to predict that the whole will work as intended even if all the parts have been fully tested and at some point you just need to give it a go; though in a controlled and measured way so that if it does not work as predicted then you will know why. If a project fails completely at the feasibility stage then you have probably not considered enough options at the start. The best insurance against failure at this stage is to ensure that you have stated your requirements completely and to have been as creative as possible at the options generation stage. Now is not the time for radically modifying the original purpose and thinking up new options because sudden changes of direction mean a lot of the effort done to this point is likely to have been wasted and this can de-motivate a team dramatically. If a show-stopper does happen it may be better to end the project, learn from the mistakes and start afresh.

It took quite a while for us to reach the point where we were ready for a feasibility test; mainly because this involved specifying, designing, writing and testing the complex computer software that would allow us to communicate digital images and other patient information securely between primary and secondary care. The details of this work are not relevant here but while this was being done the commercially funded leg ulcer research project was progressing and we were continuing to publish our work. The next major milestone was achieved on 5th November 2001 when Peter Ingham, a local GP, and I tested the secure communication link between primary and secondary care - it worked first time! We were now ready to start the formal research study just two years after the research proposal had been submitted!

The research

A formal research study is a rigorous process. Everything has to be done by-the-book and there is little or no scope for creativity once the study has started. As such it is a useful skill to acquire if you are serious about bringing about successful change; and as a team it is important that members with the appropriate skills and experience take responsibility for this phase. Pure innovators are not ideal, they get bored and frustrated by the rules; what they see as tedious attention to detail and the need to suppress their creativity. To conduct research projects well, rather then generate good ideas to be researched, you need to focus on the detail and exactly the same discipline is required when writing software. I think this is the hardest thing for an individual to do - switch from the creative and enthusiastic problem-solver to the methodical and meticulous implementer. To maintain motivation through this stage we must be organised and to celebrate all our achievements no matter how small. I believe that the reason a lot of change projects fail after the initial enthusiasm dies down is because this critical "investigation" stage of the project is either skipped or done badly. Why then it is so critical to success?

Change implies risk; the risk of failing to achieve what we set out to do and the risk of finishing worse off than we started. The fear of failure is a powerful de-motivator because failure erodes confidence and perceived threats to personal security are actively resisted. Knowledge is the antidote to the fear of the unknown and the more we know about what could happen the better we are able to make informed decisions. To keep people in the dark about planned changes is to invite failure and good communication is vital to the success of change projects. Motivational speeches have their place; but we also need the facts and we must be able to ask questions and make our minds up in our own time.

One purpose of the investigation is to reduce the risk of failure by providing information and building knowledge. Only optimists assume that no news is good news; the rest of us are more sceptical and need reassurance! Another purpose of the investigation stage is to identify the option that has the best chance of long term success for everyone: the largest win-win-win factor.

> We were well into the investigation stage and were conscious that we needed to actively work on communication with the district nurse teams. That was the stated objective after all! We did this by producing and circulating regular newsletters to everyone involved about the progress of the research project because most of them did not have the opportunity to meet face-to-face or to share knowledge and experience directly. There are always unexpected problems that arise in any research project and being open and constructive about addressing these is important. This is another reason why it is a good strategy to engage enthusiastic and open-minded early adopters at the investigative stage of a change project. They have a much greater tolerance to set-backs and are more creative in helping to find solutions to them.

The report

When an investigation comes to its conclusion it must be finished off correctly. The next new beginning, the implementation, cannot start until the previous ending is complete. For an investigation this means producing a report that contains the results, the analysis, and the new knowledge that has been gained. The report marks the end of the investigation and it should be used to celebrate the successes and provide explanations and lessons from the mistakes. I believe that projects fail for only two reasons - poor planning and unpredictable events. An unpredictable event is forgivable, poor planning is not. To be useful the report should be presented as answers to the original questions and translated into the different contexts and language needed by all the stakeholders. The knowledge gained should be presented in two ways as a summary, along with appropriate statistical analysis, and as examples. Both have their uses - the summary answers the question "why" and the examples help answer the question "how". Many people remember new knowledge better with the help of real examples and stories that illustrate the principles in action. This is why story-telling is such an effective way to communicate and why I am using our story to illustrate the general principles of successful change management in action.

Three Wins

The LUTM study was a randomised trial designed to compare the conventional paper-based method of communication between the community and hospital teams with the new method using the shared electronic record. After the community nurses had made the decision to refer for specialist assessment in the One Stop Clinic, the patients who agreed to participate were randomly assigned to one of the two groups: paper or electronic communication. This study design was intended to eliminate bias and to ensure that only the effect of using the electronic record was measured. We imposed no restrictions on the size, type or duration of the ulcers so that the study was a true representation of real life: warts and all. All patients were followed up for six months to record how many ulcers healed and how quickly, and to record the true cost of treatment in the community; the number of district nurse visits and the cost of the dressings used.

Group	n	Ulcer size at referral	Time to clinic	Healed at 12 weeks	Healed at 24 weeks
Paper	25	8.9 cm^2	41 days	38%	60%
Electronic	25	13.2 cm^2	12 days	64%	78%

The results showed that the use of the LUTM system was associated with a significant reduction in the time from referral to being seen in clinic, an improvement in healing rates at both 12 and 24 weeks. These improvements in the quality and effectiveness of care for the patients were associated with a reduction of 26% in the community cost of treatment, amounting to around £200 per patient, and it also represented a benefit to the hospital because the number of follow up visits to the hospital were reduced from five to two. This latter benefit meant that more appointments for new patients could be offered with the existing resources. We did not set out to deliberately measure staff satisfaction because by deliberately selecting early adopters for the study we would have biased the result - however we had clearly demonstrated both a quality and performance improvement. By combining the One Stop Clinic and the LUTM communication system we had effectively reduced the time from referral to diagnosis from 26 weeks to 2 weeks!

John's Story

John's story is just one of many that I could tell from the LUTM study and demonstrates very well the advantages we have experienced since using a shared electronic record as a communication tool. John's constant companion for the previous 15 years had been a leg ulcer that he had been treating himself and which had reached the size of a dinner plate before it came to the attention of his district nurse. Once she was on the case things moved a bit more quickly; John agreed to be included in the LUTM study and randomly allocated to the conventional "paper based" group. The referral letter from his GP arrived 12 weeks later in the post and he was seen in the One Stop Clinic 3 weeks after that (first arrow). The diagnosis was established, treatment started, and this was followed by a dramatic response (A) with a reduction in size by 50% from 300 cm^2 in just 5 weeks!

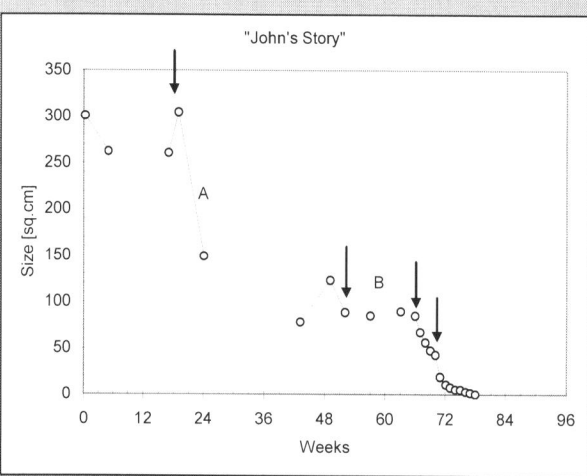

Following that initial response the healing slowed and despite being seen in clinic regularly over the next year and being offered skin grafts John refused any hospital treatment. We were able to demonstrate the lack of healing (B) quite clearly using the graph generated by the LUTM software and eventually, out of frustration, I asked John why he repeatedly refused the treatment that we knew would help. "Because I don't want to come into hospital" was the reply. My assumption had been that he was just set in his ways; in reality he was scared of hospitals! I explained that we could do the procedure in outpatients under local anaesthetic - "OK" he said and with two simple procedures (last two arrows) the remaining ulcer healed completely in less than 10 weeks! John was delighted with the outcome and I learned another valuable lesson in not making untested assumptions. We continue to follow his progress remotely and his ulcer has remained healed!

At the end of a successful investigation we should have moved along the *Diffusion of Innovation* curve because we should have convinced the early adopters who were involved in testing and evaluating the idea; and also convinced those who learn of the results and who were prepared to challenge their own assumptions. This is crucial to disseminating the new knowledge more widely in preparation for the final stage; implementation.

> At the end of the research study we were pleased with the outcome because we felt we had proven that our idea was feasible. We invited all the participants to a closedown meeting, presented the final report, thanked everyone for their help, and awarded some token "prizes" from what was left of the small research grant. I assumed that this phase of the project was now complete because we had evidence that LUTM delivered significant benefits. However, I did not believe that this would be enough to bring about a long term change and that there would still be a long journey ahead to convince anyone to adopt the new method. I was wrong because once again my prejudices led me astray. I did not understand the importance of the Tipping Point.

The tipping point

The concept of a *Tipping Point* is so obvious in hindsight that it is a useful insight to have when setting out on a change project. The first principle of change is *"Start with the end in mind"* and that also means considering how an innovation might be implemented if it turns out to be useful. The *Tipping Point* is when the idea gains sufficient acceptance and momentum that the innovators and early adopters no longer needs to push; the rest of the population start to *pull* the innovation; they want it! Malcolm Gladwell describes this principle in the context of a range of innovations and discusses the reasons why some good ideas take off and some don't. Gladwell concluded that it is not just a matter of how good the idea is (the message); success also depends on who delivers the message (the messenger) and how the message is delivered (the context). If we link this concept with the *Diffusion of Innovation* principle we can see that again it makes good sense to actively recruit the early adopters at the investigation stage of the project; by doing so we make the project easier to manage and should also move the idea towards the *Tipping Point*. If the investigation is successful then the early adopters will "sell" the idea; if not then they are the ones who are most likely to be open-minded, learn from the experience, and possibly even modify the approach and try again.

> At the end of the formal LUTM research study we had to turn off the prototype Leg Ulcer Telemedicine System as outlined in the research protocol that had been passed by the local research ethics committee (LREC). Given the successful outcome we planned to continue the development of the system at Good Hope Hospital as we were already convinced of its benefit and possibly even find a way to make it available as a routine service. LUTM was still a prototype and therefore could not just be rolled out in its current form. There was a riot! The district nurses that had been involved in the research project reacted strongly - *they wanted LUTM back* - it had become adopted and was now part of their preferred way of working. To take it away now was to threaten the benefits that they now enjoyed from improved communication between primary and secondary care. We had unwittingly demonstrated the third win - improved motivation - and although it was very rewarding to have our hard work appreciated in this way it left us with a new problem and taught us two lessons. Firstly, don't underestimate the impact that just investigating an idea can have; and secondly that when setting out to test an innovation you must consider at the start what you will do if you do, against all the odds, you turn out to have a success on your hands!

The adoption

There is a puzzle called Sudoku that became very popular in the UK when I was writing the first edition of *Three Wins*. Sudoku is a bit like noughts-and-crosses and a bit like a crossword that uses numbers instead of letters. Its attraction is that it is deceptively easy to play - it has only one rule; it requires no prior knowledge; and you can always work out the solution with enough time and effort. All you need to do is count from 1 to 9, think logically and be organised. The game is played in a square grid containing nine smaller 3 x 3 grids (squares) and the rule is simple: each row, column and 3 x 3 square must contain the numbers 1 to 9. The "problem" is presented as a partially completed grid and the "challenge" is to complete the whole grid. An individual Sudoku puzzle can be made as easy or as difficult as

Three Wins

required by how the complier designs the partial solution; by how many bank squares there are and how they are arranged.

One reason why I mention Sudoku is that it is a perfect example of *Diffusion of Innovation* and *Tipping Point* and it is worth asking why this new puzzle became adopted so widely and so quickly. Firstly, it is challenging enough to be interesting and you know at the start that there is a solution. There is motivation to try because you *believe* that success is possible and you can achieve it. You get a sense of personal achievement from solving the puzzle. Secondly, the challenge has been made widely known by being published in the puzzle section of daily newspapers and has therefore come to the attention of a large number of people and therefore enough *early adopters* curious enough to give it a try. Thirdly, the challenge is not too difficult or too dependent on special skills and so is accessible to a lot of people who by demonstrating early wins on the easier puzzles become motivated to take on harder and harder ones. Sudoku has appeal to a global population who may give it a try, become motivated and even addicted, and then promote the spread of the innovation. These are the three factors that Malcolm Gladwell states are needed to achieve the *Tipping Point*: the right message, the right messenger and the right context.

		3	1	4	6			
	1				7	5		
	4			2				1
	7				8	9		
3		9	7		5	4		
	9	6		5				8
	3							
4			1	7				9

8	7	9	3	5	1	4	6	2
2	6	1	4	8	9	3	7	5
5	4	3	7	6	2	9	8	1
6	5	7	1	2	4	8	9	3
3	1	2	9	7	8	5	4	6
9	8	4	5	3	6	2	1	7
7	9	6	2	4	5	1	3	8
1	3	5	8	9	7	6	2	4
4	2	8	6	1	3	7	5	9

The left panel shows the Sudoku "problem" and the first step in the solution - the only value between 1 and 9 that satisfies the rules for the rows, columns and square. The panel on the right shows the completed solution (original problem shaded). This example is an Easy one.

Three Wins

> I was introduced to Sudoku by my wife Abigail who had come across it in a daily newspaper. When I looked at the example she showed me I exclaimed "This is a simple problem for a computer. Why waste lots of time solving it yourself when you can get a computer to do it in a fraction of the time?" and promptly disappeared into my office to write a program to solve Sudoku puzzles. To me the challenge was academic; to solve the problem once and for all. To Abigail the challenge was pragmatic; to be able to solve a puzzle herself and gain a sense of achievement from doing so. Later, after I had written a program to solve any Sudoku puzzle of any complexity, I too found I got personal satisfaction from spending time solving Sudoku problems by hand; even though I knew it would be quicker to use the computer. It occurred to me that this behaviour may have parallels with the way we manage recurring problems in healthcare. If you get satisfaction from solving the same challenging problem over and over again then what motivation is there to solve the problem once and for all? Could it be that we have become addicted to the pleasure gained from fire-fighting the problems that we ourselves have created? If so why not get the same buzz from finding the root cause and fixing the problem once and for all; and re-invest the time freed up in the next challenge?

There is another very good example of the application of the *Diffusion of Innovation* but this time in the NHS. Led by Sir John Oldham, the National Primary Care Development Team (NPCDT) implemented widespread change across GP surgeries in England in a remarkably short period of time. Sir John's book tells a powerful story of how the principles were applied successfully using the Collaborative Model espoused by Don Berwick of the Institute for Healthcare Improvement in America. The explicit objective of the NPCDT was to reach the *Tipping Point* where approximately 20% of GP practices had adopted the innovation - from there the momentum of change would be enough to complete the diffusion of new ideas. According to Sir John these principles were well known to the Romans and the translation of the title of Sir John's book is *"Every system delivers exactly the results it is designed to give"*. The conclusion I drew from this matched my own experience: "If your system does not give the results you want you need to re-design it" and re-designing complex healthcare systems to deliver successful win-win-win solutions is not easy. However, it is possible; I know because we achieved it.

There are other reasons why Sudoku is a useful metaphor and very relevant to this book:

1. Sudoku is a good example of a problem where each part of the solution has to meet several simultaneous objectives to be acceptable (win-win-win).

2. Sudoku is a good example of a problem where some bits are easy (low hanging fruit) and some are not (high hanging fruit).

3. Sudoku is a good example of a problem where persistence, innovation, and rigorous application of the "rules" will reveal the safe path to the solution.

In other words, solving a Sudoku puzzle is a useful analogy to solving complex healthcare service improvement design problems!

The *theorists* prefer to prove that the general problem can be solved and to develop the theory of how to do it; the *pragmatists* prefer to apply the proven theory and solve the specific problems in reality. Both get pleasure from solving the problem and both are needed but they are as different as chalk and cheese.

What I learned is that the combination of problem-solving in theory and in practice is better; the sum is greater than the parts.

Three Wins

Chapter 5. Implementation

The final stage of a change project is to implement the option that the investigation has identified as the "best" - in other words to diffuse the innovation to the early and late majority groups and even the die-hards! William Bridges in his book *"Managing Transitions"* concentrates on this phase of a change process and quite rightly so; just having a proven idea is no guarantee of that the change will be implemented successfully. The more work that is done in the earlier stages and the deeper the understanding of the challenges that lay ahead the greater the chance there is of success. Just as the innovation and investigation phases require specific skills and roles, so does the implementation phase. The emphasis of this phase is to manage the human aspects of the change at the same time as the specific detail of the project implementation and Meredith Belbin observed two characteristic team roles that were particularly valuable during implementation; the *Team Worker* and the *Implementer* (formerly called the Company Worker). These two roles ensured that the human aspects were handled sensitively and all the necessary tasks were done according to the plan. It is at the implementation stage when the detailed project plan and a business case are needed and these must both be based on the evidence that comes from the innovation and investigation stages; because without this knowledge the financial and practical risks cannot be accurately assessed. The other important element of the implementation stage is the communication plan; and Bridges highlights this as a potential showstopper if it is either forgotten completely or left until the last minute. The implementation is a single path that consists of a sequence of actions that must usually be completed in a specific order. The skill of planning and managing this stage of the project is to ensure that everything gets done to specification, on time, within budget and with the active cooperation everyone affected by the change. Some tasks are more important than others and some things have do be done in a strict sequence; effective communication is critical at all stages. It is this implementation phase that most people would equate with project management and it is important not to confuse innovation and implementation; they are quite different and require different tools, skills and methods. Innovation and implementation are synergistic and you need both but they are also as different as chalk and cheese.

Three Wins

> The first part of our story is almost complete and, following the distress caused by turning off the prototype LUTM system, we did achieve our goal by completing the implementation stage. The district nurses, with the support of their GP's, canvassed their Primary Care Trust (PCT) managers to consider a formal implementation of the LUTM system for all the community nursing teams in North Birmingham. We produced a project plan, training plan and business case to support the implementation; the software development was completed to a commercial standard and made available through a formal software licensing agreement. The implementation included the cooperation of a large number of people at Good Hope and in the PCT and was coordinated using conventional project management techniques such as controlled documents with a small number of well-focussed meetings. The full implementation of the system across all the community nursing teams took only six months and has since been extended to two other local PCTs that refer patients to the Vascular Surgery One Stop Clinics. The one-off cost of the full implementation of the LUTM service amounted to about £0.50 per head of population which, given the cost savings that we had demonstrated in the LUTM trial, compares favourably with cost of leg ulcer care in the UK which is now about £20.00 per capita per year.

The path

A technique that is very useful at the implementation stage of a change project is called *critical path analysis* which is a formal method that allows you to identify the implementation plan steps that are most sensitive to failure. There are two parameters of each step that need to be considered - the dependency and the criticality. Dependency is the requirement that one step must follow another, and criticality is the requirement that a specific step must be completed at some point. There are a several basic principles that are useful in planning project implementation paths and these are

> (a) keep paths as simple as possible,
> (b) keep separate tasks independent of each other if possible
> (c) organise tasks to run in parallel if possible, and
> (d) minimise the risk of failure of the critical steps.

The use of contingency options is helpful here because it makes explicit what possible actions are available *if* a critical step failed and were to threaten the whole project. In general the less dependency there is between tasks the more parallelism should be used because parallel working allows projects to be done more quickly provided you have enough resources and provides more resilience in case of failure. The greater the dependency between tasks then more sequential a path must be or the lower the risk of failure that can be tolerated. The critical path is the one that determines the minimum time in which the project can be completed. A poor combination is high dependency and high risk of failure as this equates to a lower chance of success.

Different options represent different parallel paths and it makes good sense to prioritise tasks that appear in more than one option; this way you have a greater chance of being able to switch to a different option without back-tracking or wasting effort and this makes your project more resilient and more likely to succeed. This well known strategy is called *covering your options* and once again this emphasises the importance of spending time at the innovation stage in generating lots of options and time at the investigation stage testing them!

> On reflection I realised that we had intuitively rather than explicitly managed the critical path. We did things that we could do, identified things we needed to but couldn't yet do, made the best use of opportunities and avoided predictable pitfalls. These skills were probably acquired from a number of sources but most likely from years of clinical experience. Patients with complex, chronic disease are predictable in general but not individually; and although the principles of management are clear, the skill is applying those principles in practice. Healthcare professionals are not in the business of taking unjustified risks; the outcome is not completely predictable, unexpected adverse events happen; the challenge is to minimise the opportunity for adverse outcomes and maximise the likelihood of the desired outcome. Healthcare professionals already have the skills needed to do this for unhealthy people, and based on our experience they also have the skills needed to do it for "unhealthy" processes.

The finish

The end of the implementation stage is the time to take stock and to compare what we actually achieved with what we intended to achieve; to compare the deliverable with the original requirements. This should be done as objectively as possible using the measures that were identified at the start. We are now at the last stage of the design process and looking back to the beginning. The risk at this stage is to fail to achieve closure and this is where the Belbin team role of *Completer-Finisher* comes into its own again. It is important to ensure that all the loose ends are tidied up; jobs like completing and distributing the project report, finalising the finances, and filing the paperwork. It is surprising how after change projects have finished, there is often no permanent record of all the work that went in to achieving the success and therefore little to pass on other than anecdote and experience! Without completing properly it is much more difficult for the lessons, both good and bad, to be learned and shared. I am sure that there are many examples of successfully implemented innovations that remain isolated and probably re-invented over and over again because there is no easy way to disseminate the new knowledge. It is for this reason that I am now wearing my *Completer-Finisher* hat and reviewing what we did with the benefit of hindsight. My motivation is to learn and to avoid repeating my own mistakes in the future! I am trying to *Walk- the-Talk!*

The final twist in the first part of our journey happened unexpectedly. One day early in 2004 I was approached by David Gleaves of the West Midlands NHS Innovation Hub who had heard of our work and who suggested we submit the project for the annual NHS Innovation Awards. It was never one of our original objectives to do any more than publish our findings - we certainly didn't do it to win any awards! After all there must be dozens of well-funded and well-managed research teams that would have much more impressive projects. However, I took David's advice and a few weeks later was surprised to receive an invitation to attend the *NHS Live* event in London with the other runners up. It was a nail-biting finale and I was surprised and delighted to accept the first NHS Innovation Award for Innovative Service Delivery on behalf of the whole team! That was July 2004. Once again this lesson illustrates how we all make false assumptions that create barriers to success and how opportunities often present themselves when least expected. It was at the *NHS Live* event that I was asked the deceptively simple question "How did you do it?" It was that question started me on the path to writing the first edition of *Three Wins* and to continue to developing the thinking and methods that I now call Value Stream Design.

Chapter 6. Complexity

At this point in our story you might be experiencing some disbelief, some frustration, and even some disappointment! You might be saying to yourself "*Yes, well that all sounds easy enough for you and the well defined problem you had to solve but the reality of my situation is much more complicated than that.*" You are right. The problem we had to solve was well defined and the principles I have outlined are simplifications that are easy enough to see working in hindsight but much more difficult to apply with foresight. We too came to appreciate the challenge of complexity as you will soon see. So, to address the challenge we will just apply the principles we have learned: When you don't know what to do, start doing something, start asking questions, questions like "What, where, when, who, how and why?"

> Q: Why does reality appear more complicated than this example?
>
> A: In real situations there are lots of things happening at the same time and although we can see how we might apply the principles to one problem we find it very difficult to keep track of all the problems when all the problems are interconnected.

This is the essence of complexity; lots of *inter-dependent* processes running at the same time. How often do we think to ourselves "I wish everything else would go on hold while I concentrate on this job because all the interruptions are stopping me getting anything done!" That is the problem with complexity; we can't get away with just fixing one bit at a time; we also have to manage the complexity. And it appears that we can't "think" our way logically and linearly to the solution; we only seem to be able to "feel" our way and work it out as we go.

Just as with the Belbin team roles, we each have a preferred way of thinking and this "going with the flow" style is more comfortable for some than others. To surf the wave of change requires flexibility and an ability to think in different ways as and when needed. It is not that one way is better than another – you need all the tools in the box and know which tool to use for which job. Learning to think outside our comfort zones is uncomfortable at first but it gets easier with practice.

The game

Remember the Sudoku puzzle? It has a single rule: "Each row, column and square must contain the numbers 1 to 9" and there are 9 rows, 9 columns and 9 squares that all need to apply this rule *at the same time*. The Suduko puzzle is 27 simultaneous interdependent problems that are interacting with each other; a 27-ball juggling trick! The way to solve a Sudoku puzzle is just to apply the rules and do the bits that are obvious first because this makes the rest of the puzzle easier to solve. It is the *Divide-and-Conquer* and the *Low-Hanging-Fruit* principles working together and what makes Sudoku puzzles challenging is not the principle of how to solve them - it is the practice. Rather like caring for sick patients and working with teams to heal "sick" healthcare processes! As individuals we are not very good at working on multiple simultaneous linked problems like this because when switching from one bit of the puzzle to another we tend to forget where we were and we make mistakes. The rule may be simple enough to understand but applying the rule without making mistakes is not so easy.

OK, so if it is difficult, slow and error prone for one person then let us divide up the work and put a team of 27 people on the job; ask one person to concentrate on just applying the rule to each row, column, and square and ask them to only look at their part of the problem; all they have to do is apply the rule and say if a proposed solution is acceptable to them. Everyone has a clear understanding of the why - to get a sense of collective and personal achievement; the what - to solve their bit of the puzzle; and how - to apply the rule. All that is needed now is a coordinator and effective communication between the coordinator and the team members. The coordinator looks at the whole picture, suggests options, and records the replies from the respective row/column/square members of the team. The coordinator just asks questions "Does this option satisfy the row/column/square rules?" In other words is this option a potential win-win-win solution? If not then that option is rejected. Of course, while the solution is incomplete some squares may have more that one possible win-win-win option; others have only one. Finding the squares with only one option means the answer for that square is defined and the rest of the puzzle has become a bit easier. You progressively move closer to the final solution which is a win-win-win for every cell in the grid. The innovation stage is deciding the possible options to propose; the investigation stage is applying the rules; and the implementation stage is completing the grid.

In theory this 28 person Suduko team would work but in practice it would be a very inefficient use of resources because most of the time the team members will not be doing anything. One alternative would be to put one team member on rows, one on columns and one on squares and have the fourth person acting as coordinator - a team of four is easier to manage and this way all the members will be working all the time and would get the solution just as quickly. In Sudoku the only way to fail is by making mistakes or false assumptions and to go up a blind path. With Sudoku you don't get the wrong answer; you don't get an answer at all and that is how to waste a lot of time end effort getting nowhere. There are no rewards for not finishing; and trying to backtrack on a failed attempt is more difficult than getting it right first time. It seems to be this risk of failure that creates the excitement and the sense of achievement when the problem is solved.

In contrast to Sudoku, the simpler game of noughts-and-crosses is not emotionally satisfying. Noughts-and-crosses it is not challenging enough; and it is also a battle with only one possible winner. When the players are equally matched the outcome is always a stalemate. All the game serves to illustrate is the futility of conflict and the time, effort and money that conflict wastes.

> We had several steps in our journey to designing a process that delivered higher quality care; that was more productive; and that provided a more motivating environment for staff. The requirements for each step were clear and we did the easier bits first which made the rest of the problem more tractable. We started with an easy win and worked our way through the more difficult and challenging research and implementation stages towards our intended goal, changing the plan as opportunities or obstacles appeared. We pragmatically solved the puzzle but we had to do it sequentially. We had no training, guidance or external support and it took a long time!

The paradigms

The problem with complexity is that as individuals we seem to be limited to how many mental balls we can juggle at the same time. With practice you get better but there still comes a point where even the most accomplished mental juggler starts to drop things. There is a point when the problem becomes too big for one person to have a complete understanding; the complexity limit. When faced with this situation we appear to adopt different strategies:

1. Avoid problems that we can't understand fully.
2. Simplify the problem to the point where we do understand it.
3. Divide the problem up into smaller bits that we can understand.

The risk with Option 2 is that oversimplification can lead to confusion when the real system does not behave the way that our simplified version predicts! Option 3 is the application of the *Divide-and-Conquer* principle and the assignment of different roles and responsibilities for parts of problem to different people with different views and skills. Each person has a simplified but manageable view of the bigger problem and by coordinating their efforts a path to a solution can be found; just as with the Sudoku puzzle.

Each simplified view of the problem is called a paradigm or a worldview; and the difficulty with the *Divide-and-Conquer* approach is that each person has a different paradigm and they only make progress towards the collective goal when all agree that that a proposed option is acceptable. The same is true of real world problems; so often we seem to argue about something we actually agree on; we seem to spend a lot of time in *heated agreement*. The solution is to get information flowing effectively through the team, most importantly between the coordinator and the implementers. To do this means translating the problem into different forms of language; forms that are more suited to the recipient than the sender. Do not expect the implementers with their different paradigms to be able to easily communicate directly with each other to solve the problem. In the four-person team working to solve the Sudoku problem one sees the problem as rows, another sees it as columns, another sees it as squares; the fourth person has to be able to speak all three "languages" in order to maintain progress. Only when all four are communicating effectively does the path become clear and the problem can be solved.

Three Wins

The Blind Men and the Elephant Story
(adapted from the poem by John Godfrey Saxe)

"Three blind men were discussing exactly what they believed an elephant to be, since each had heard how strange the creature was, yet none had ever seen one before. So the blind men agreed to find an elephant and discover what the animal was really like. It didn't take the blind men long to find an elephant at a nearby market. The first blind man approached the animal and felt the elephant's firm flat side. "It seems to me that an elephant is just like a wall," he said to his friends. The second blind man reached out and touched one of the elephant's tusks. "No, this is round and smooth and sharp - an elephant is like a spear." Intrigued, the third blind man stepped up to the elephant and touched its trunk. "Well, I can't agree with either of you; I feel a squirming writhing thing - surely an elephant is just like a snake." All three blind men continued to argue, based on their own individual experiences, as to what they thought an elephant was like. It was an argument that they were never able to resolve. Each of them was concerned only with their own experience. None of them could see the full picture, and none could appreciate any of the other points of view. Each man saw the elephant as something quite different, and while each blind man was correct they could not agree."

The message in the story is that the whole was more than the sum of the parts and although the parts were correctly described each was a simplification. Understanding the whole together requires the ability to view the whole from many perspectives and this means challenging the assumption that your view is complete or the only one that matters.

In our project we had many stakeholders, each with their own perspective that we needed to appreciate. We had to consider the views of the patients, the outpatient staff, the community nurses, the hospital managers, the PCT managers, the IT department, the R&D department, etc. We needed to find a path to the shared vision that was acceptable to everyone at all times; and that meant considering all the possible options, asking lots of "Do you have any good reasons why we should not do ...?" questions, and doing the obvious, agreed and easy bits first. This strategy worked for us and meant that our limited resources were focussed on just what needed to be done and what could be done at each stage. With more resources we may have done more of the tasks in parallel and reached the goal more quickly; we may also have lost our focus, got lost, failed and given up.

The constraints

So far I have concentrated on the principles and tools that help create and maintain the critical factor in finding and implementing win-win-win solutions: motivation. Improved motivation is also one of the wins so we must maintain the balance and not neglect the other two wins: improving the quality and improving the performance of the service. Unlike the one-rule game of Sudoku each of our three wins has different rules and different constraints; but the principle of how to find a solution is the same; only accept an option that meets all the constraints. However, now the problem has got a bit more complicated. How do we apply quality and performance constraints? What are the rules that we need to use? To answer this question we need to define what we mean by the words *quality* and *performance* and we need to do this separately for each of the three views; the patient, the staff and the organisation. Constraints can be looked at two ways; what you are not prepared to accept (exclusion) and what you must have (inclusion). Sometimes it is just simpler to use one or the other; though often you need to use both.

These two views of the win-win-win objectives of quality, motivation and performance define a standard for quality as "No mistakes", a standard for motivation as "No threats" and a standard for performance as "No waste":

Goal	Paradigm	Objective	Standard
Win	Patient	Quality	"No mistakes"
Win	Staff	Motivation	"No threats"
Win	Organisation	Performance	"No waste"

The challenge is to find a set of options that meet all these constraints at the same time and then to implement the easiest, cheapest, least risk option. I suspect that if I had posed this challenge at the start you would say "That is impossible!" but by now I am hoping you are saying "That *might* be possible"

It is possible if you focus on achieving a win-win-win outcome and apply to the principles of value stream design.

The solution

It is only when the final solution is implemented that you will know for sure if you have achieved your objectives. The solution is the combination of the people and the processes; just as the solution to the Sudoku puzzle involves both the player and the rules. For the solution to work all the components of it must work because everything is interconnected. I regard a solution as *competent* if it is capable of meeting the constraints. A mistake in one part of the solution may have a knock-on effect that can affect the whole solution; so all the parts must also be *competent* in their separate but related roles. In other words the staff must be competent to complete their assigned tasks and the processes need to be competent to provide the appropriate context for this to happen. Ensuring that the staff are competent requires selection, training, appraisal and support; ensuring the processes are competent requires *designing* them to be so.

A digital computer is a good example of this principle in action. The central processing unit (the "brain") is the dynamic component that actually does the work; the program (the "instructions") is the static definition of the process. Without a program the CPU will be idle - twiddling its digital thumbs – but still using resources (electricity) and still getting warm; a rather expensive heater! All the creative effort goes into defining, building and testing the program (the process) but without the CPU this creative effort is also wasted. Only when the CPU is combined with the program does the electricity get converted into something more useful than heat. Similarly, only when competent staff and competent processes are combined does the time and money spent produce the desired outcome; and the better they work together the less time and money is required. Quality and performance only happen when the staff and the processes are competent and synergistic; the sum is greater than the parts. They are inter-dependent.

In my experience it is not the staff that are "broken" - it is the processes that are broken. The staff are just trying to do their best within the broken process. The problem is that the staff get emotionally damaged by the broken process and signs of that damage are manifest as de-motivation, cynicism, anger, aggression, blame and a range of other unconstructive behaviours that only serve to make the working environment increasingly toxic for everyone else; what I call the *Toxic-Emotional-Waste* syndrome.

Three Wins

> After we had completed the Leg Ulcer Telemedicine Project a new and unexpected problem developed. We became the victim of our own success as we experienced a progressive increase in the number and the complexity of leg ulcer patients referred to the One Stop Clinic. If we had been a commercial organisation this might have been a welcome opportunity; but for a resource-constrained healthcare service it was a threat. The win we had achieved for the staff was now being eroded as clinics were becoming over-booked with complex new patients and the waiting times for all patients were starting to increase. By demonstrating that a solution was possible we had uncovered an unmet need and to sustain the benefit we now needed to increase the performance of the outpatient clinic without compromising the quality of the service. Once again it was not a question of what we were doing but how we were doing it and this meant starting a third cycle of change. We had implemented the one stop clinic (low hanging fruit); we had implemented the shared electronic record (difficult to reach fruit); we now needed to design new processes that would improve the performance of the clinic at no extra cost but at the same time maintain the service quality and staff motivation we had worked so hard to achieve (a juicy piece of high hanging fruit!). So back to the drawing board; back to Stage 1 of the design process; and back to asking questions: "Why did an increase in demand cause this new problem?" The answer was that we had not changed the way in which the clinic was booked. "Why?" Because we didn't think we needed to and because it involved another department who showed little motivation to change; so quite sensibly we did not put this it on our critical path. We were alright up until now but we had changed the context by implementing the LUTM service and the reason it had now become a problem was because in changing to a One Stop Clinic we had deliberately increased the variation between the patients. Some patients just needed 15 minutes with a doctor, others took up to 90 minutes to go through the sequence of nurse-led assessment, diagnostic tests, clinical review and then treatment. The One Stop Clinic was a win for the patient but had exposed a limitation in the organisation: the booking system was not designed to work this way.

The model

There is another way of solving a Sudoku puzzle - get a machine to help with the tedious and boring work of applying the rules and identifying options that *fail* to meet the constraints. What is left when all the incorrect options have been eliminated is the correct solution. The reason that it is possible to get a machine to do all the work is because a Sudoku puzzle is called *deterministic*; there is single rule that when applied rigorously and without making mistakes will determine if you have a valid option. In other words you can work out if an option is unacceptable and exclude it from further consideration – it requires no creativity and no judgement. The problem with solving a Sudoku puzzle this way is there is one correct solution and very many incorrect solutions for each puzzle; millions of options that need to be tested and excluded. How can a machine be designed and built to do this? The answer is to copy the Sudoku solution process that a person would use and translate the process into a language that a machine can use. One machine that is good at this sort of problem is the humble desktop computer. The computer program represents the process so the task is now just one of designing a computer program to model the Sudoku solution process. It is the combination of the computer (processor) and the program (process) that creates a Sudoku Solving Machine that provides the correct solution quickly and without making mistakes. The creative effort goes into designing the program and after that it is just a matter of letting the CPU (processor) do what it does best - follow instructions.

A computer is quite competent to solve any Sudoku puzzle and even a basic desktop computer can find the solution in a fraction of a second! Humbling eh?

The analogy is important: it is the combination of a competent process and a competent processor that creates the desired outcome; high performance (no waste) and high quality (no mistakes). The similarity to healthcare is clear; the difference with healthcare is that it is much less deterministic; there is no single or even a small set of rules that can be rigorously applied to achieve the required outcome. The staff delivering a healthcare care service must employ their skills, draw on their experience, and exercise their judgement at every step of the process in order to avoid mistakes. A healthcare process cannot be run like a mass production line with only low-skilled workers or robots that are only responsible for a small part of the solution; the problem is just too variable. So although a machine cannot replace the healthcare worker the requirement for competent processes is just the same. Healthcare processes must be *designed* to work; they must support the staff not damage them; and they must be proven to be fit-

Three Wins

for-purpose before being implemented. The challenge is how do you design a healthcare process to be fit-for-purpose?

> What we needed was a way for the patients to be offered One Stop Clinic appointments in a way that matched the complexity of their problem and the availability of the resources. In other words we needed a clinic booking system designed around the patient care pathway. Existing patient administration systems (PAS) are designed primarily for the convenience of the hospital outpatient department rather than the patient and do not support this more sophisticated patient-centred process requirement. The problem from the PAS perspective is difficult because different patients need different resources within the clinic for different periods of time. Not only did the patients need to be booked at an appropriate time but an appropriate number of patients of different types needed to be seen to ensure that waiting times for different types of patient do not increase disproportionately. Finally, clinics must be booked so that specific resources such as diagnostic tests were not overloaded and then cause further delays for patients and staff. This was turning out to be a major challenge and the time had come to brainstorm some innovative options. We needed a more sophisticated booking system but one that could be easily implemented using the existing PAS; a simple and robust process that the booking clerks could follow easily without making mistakes. It needed to be no more complicated than the existing processes - and it had to be simple enough to do on paper. In other words the complexity of the booking problem needed to be hidden in the design of the paper-based booking template. However, there are literally thousands of different ways of booking a clinic and finding the one that would work best was like solving a highly complex Sudoku puzzle for every clinic! I needed to simplify the problem and the breakthrough came when I realised that I didn't need to solve the puzzle for each individual clinic - just for a typical clinic with the correct ratio of simple and complex problems. We knew what this ratio was because we had been collecting this data for some time. What I actually needed to be able to do was to test a number of possible clinic booking options to find one that would satisfy the known constraints. It was clearly not practical to do this using PDSA in the actual clinic - there were just too many options. I needed a simple, quick and objective way of testing lots of options and eliminating the ones that would not work. I needed a model of the clinic process to use to test my options.

The simulation

When applying this principle to assessing possible options for designing competent processes we find that healthcare processes are also deterministic. There are rules that determine how the process works; but the rules are not precise - they are fuzzy. There is variation everywhere. However, we can still use a model of the process to represent and test proposed options and we can use this model to *simulate* the process and predict how it is likely to behave. The difference from solving the Sudoku puzzle is that it is now not easy for a computer to find the solution itself; the rules for healthcare processes are too numerous and too fuzzy. The computer would have to be creative - it isn't - for creativity we need to employ the "wetware" between our ears. Enter the value stream designer.

Process simulation using computer models is a standard technique that has been used in industry and academia for many years, including healthcare operational research. Surprisingly, it has not made much impact in frontline healthcare service improvement. There are probably many reasons for this but I suspect that it is because operational research is a largely academic discipline that is not closely integrated into the operational workings of healthcare organisations. Consequently there is not enough diffusion of the known theory to the frontline; not enough examples of successful use in healthcare; not enough people in healthcare who are aware that the techniques even exist; not enough simulation tools designed specifically for patient-centred healthcare process improvement or for healthcare staff to use; and not enough appropriate infrastructure, training or support for the use of such tools in healthcare. Once again that no one has successfully applied the well-known theory in practice and spread the word.

To solve the outpatient booking template problem I borrowed a well described method from computer science and operational research called Discrete Event Simulation (DES). In DES the process is defined as a sequence of actions that occur in a specific sequence, rather like following a recipe. The time taken to complete each step must be known reasonably accurately and using this information the time when the next step in the process will happen can be predicted. This is actually easy to do on paper for a single patient; and with a board and counters for a small number of patients; but it becomes increasingly more difficult when there are many patients following different pathways and competing for the same resources (e.g. chairs, rooms, doctors, nurses, diagnostic tests, beds etc). However, the rules for running a process simulation game are simple and can be delegated to a computer which can take over all the tedious work. This means that the DES method is ideal for representing even complex systems built from a large number of simpler but interrelated processes and most importantly a DES simulation can predict how the whole complex system will behave under any conditions as well as allow detailed analysis of parts of the system. DES is a tool that enables you to handle the complexity. The healthcare value stream designer just needs to set the constraints, suggest a possible option to test and the computer uses the DES program to predict how well that option will work. In this way a range of options can be compared. The designer is the creative, coordinating half of the partnership, the computer the obedient, tireless worker. Knowing that DES was potentially one way to solve this problem, and not wanting to re-invent the wheel, I considered using one of the commercially available DES software tools. However, I quickly found that they were highly complex pieces of software to use, not very easy to apply to my specific problem and were prohibitively expensive given my non existent budget! It was a repeat of the LUTM software challenge. Fortunately I already had experience of using DES and some years before had written some software that used the method to solve a completely different (non medical) problem and I was able to adapt that software to create a patient-centred process model of the One Stop Clinic. The steps required to do this are essentially identical to the Seven Stage design process and interestingly I found that most of the components required to build a value stream process model are already widely used in healthcare. The most useful technique is mapping patient-centred processes as described by Sarah Fraser in her book "The Patient's Journey". In effect I created a tool to analyse why the existing clinic was struggling, to test novel booking schedules and to find one that was predicted to work better: I only needed one option to implement.

A process simulation can not only be used to predict the acceptability and performance of the proposed solution it can also be used to predict the cost of the solution and therefore show if the solution will also meet financial constraints. This means that provided the other factors that determine the quality of the service are also met (i.e. staff competence) then the whole quality-performance problem can be addressed from start to finish.

Healthcare process simulation provides an alternative way of investigating options for certain types of problem and like all the other investigation methods (e.g. a formal research trial) it requires appropriate training, experience, tools and guidance to apply correctly.

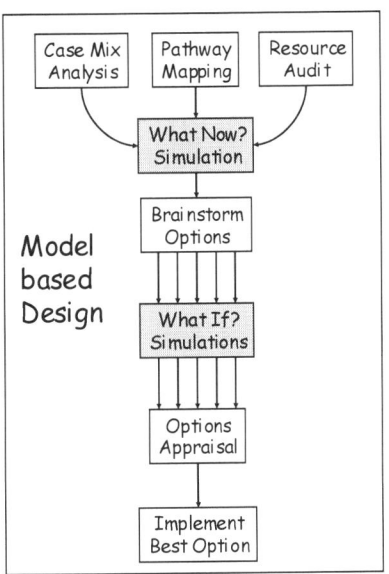 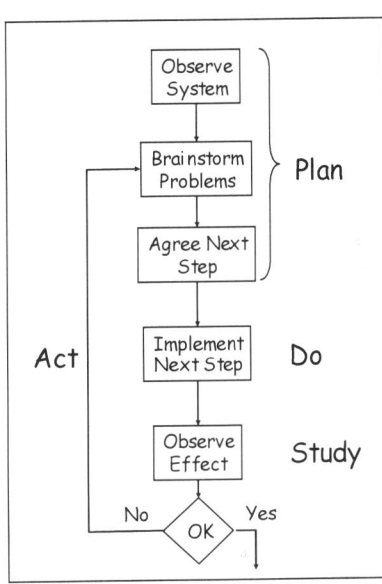

Comparison of conventional incremental improvement using the Plan-Do-Study-Act (right) with single step simulation-based process redesign method (left). These two methods are synergistic as the model-based approach can be incorporated into the planning stage of a higher level PDSA strategy and PDSA can be used to manage the fine detail at the tactical level of best option implementation.

A process simulation is just one of many ways to investigate the feasibility of a proposed option and to use process modelling appropriately it is important to know its strengths and weaknesses. The main strength is the ability to accurately predict the future performance of a complex system of interrelated parts; the main weakness is that it requires specific skills, tools and experience and if these are in short supply then process modelling should be limited to problems that justify the extra effort required. That said, the only way to acquire the necessary experience in using process modelling is to learn the basic skills; develop those skills and expertise on well known problems and only move onto the more complex ones when the basic skills are mastered. Alternative service improvement methods such as PDSA and DES are synergistic and they should be used together rather than be seen as competitors.

Process simulation has another application; as an educational tool. Simulations allow students to interact with the process model and observe the effect of making changes on the whole system in a safe and controlled way. This role of process simulation enables healthcare staff to acquire the necessary skills for competent service redesign irrespective of whether they use process simulation tools or PDSA or both. For example, just having an awareness of the impact of patient case mix variation on the number of beds required to avoid day-of-surgery cancellations helps in the day-to-day running of a hospital. I don't believe we should aspire to use sophisticated simulation tools to run a health service; I have learned from experience that we can use these skills and tools to design a health service that will run smoothly on its own.

> The output of the One Stop Clinic DES design exercise was a paper-based booking template (see below) that predicted an increase clinic maximum capacity of 40% (from 16 to 22 patients) without the clinic running late or burning out the staff in the process. The increased capacity within existing resources was created by ensuring that patient delays within the clinic were eliminated as much possible by designing the schedule so that the resource each patient needed next became available just-in-time. This meant that critical resources would be busy most of the time if the clinic was fully booked but the clinic would still finish on time. From past activity data we knew that the average number of patients seen in a clinic was 16 so the maximum booking template with 22 slots allowed sufficient resilience to cope with peaks in demand and to reduce any existing waiting lists. The new template was easy to implement because the process for the booking clerks was easier than the existing computer-based system of complex booking rules. Once the booking clerks had direct experience of the benefit of the designed schedule they became surprisingly motivated to change the way they worked; just as we had observed with the community nurses and the LUTM system. I know this because of another unexpected event that occurred after we had implemented the new booking template. When we piloted the new booking template in

> my clinic I was holding One Stop Clinics in the afternoon from 2 to 5 pm and my colleagues ran theirs at other times of the week. For unrelated reasons I had to swap one afternoon clinic with a colleague who had an equivalent morning clinic; a straight swap so no problems were anticipated. It was quickly apparent that this was not the case - the new afternoon clinic regularly ran over time and caused the clinic staff to start complaining. What had happened? Further enquiry revealed the cause; before the swap the morning clinic (that was not using a patient-centred booking template) regularly over ran but this did not cause complaints because the staff had learned to accept that on that day they would usually not get a lunch break. However, the same staff would not tolerate the clinic running after 5 pm because this caused domestic problems! Interestingly it had been this emerging trend with my afternoon clinic that prompted me to design the new booking template. The initial response to this new and unexpected problem was to reduce the number of clinic slots in the afternoon (i.e. reduce the maximum capacity) and the more measured response was to use the new booking template to restore the original capacity without incurring complaints from the staff.

This experience illustrates very well a number of the lessons that are recurring through this book; most important of which is that even adopting proven innovation throughout a single department is not necessarily automatic or easy.

Our experience reinforced the previous conclusions:

Accepting ownership of the problem is the first step to finding a solution;

Believing that a solution exists is the second step; and having the

Courage to make the commitment to make the changes is the third.

Three Wins

		Mr S.R.Dodds - One Stop Outpatient Clinic GHH - Booking Template								
Day	Monday					Instructions: 1. Use the category assigned on the GP letter to book into the NEXT free slot of that type. 2. ONLY book urgent patients into urgent slots. 3. Ensure the templates folder is taken to all clinics and returned to the booking office after the clinic.				
Time	09:00 - 12:00									
Date										
Version 06/01/2005										
#	Time	Patient Type	Number	Surname	Forename	DOB	Cons	CNS	V-Lab	Dress
1	09:00	New Leg Ulcer						09:00	10:00	10:30
2	09:00	FU Leg Ulcer								09:00
3	09:00	FU Leg Ulcer								09:00
4	09:00	New Vascular					09:00	09:15		
5	09:15	New Vascular					09:15	09:30		
6	09:30	New Leg Ulcer						09:30	10:30	11:00
7	09:30	FU Leg Ulcer								09:30
8	09:30	New Vascular					09:30	09:45		

… to 22 patients in total.

Example of the first part of the One Stop Clinic booking template designed and tested using the discrete event simulation method. Patients are booked according to their clinical category to arrive at the starting time on the left and are scheduled into the different resource streams of the clinic as indicated by the columns on the right. This timetable ensures that patients experience minimum delays and the resources are neither under-utilised nor over-loaded. Patients are booked first-come-first-served into slots of the appropriate casemix type and the whole process is done on paper without the need for complex clinic scheduling rules or software. As well as creating the context for a higher quality and higher performance service it ensures that the activity is recorded accurately to ensure the greater complexity of the One Stop Clinic is properly reimbursed.

Chapter 7. Value Stream Design

The term Value Stream Design (VSD) describes what it is; the *value* is the care the patient wants; the *stream* is the flow of patients, work, information, expenses and revenue; and the *design* is the combination of creativity and objective effort to build processes that are fit-for-purpose. The measures of success in VSD are the Three Wins: higher quality; better performance; and improved staff motivation. Since publishing the first edition of *Three Wins* I have invested considerable time reading the teachings of the schools of *Lean Thinking*; *Six Sigma* and *Theory of Constraints*. The good news is that we are all in agreement; it is the same set of principles, the same message just repeated in slightly different languages. To illustrate this I will attempt to translate the *Three Wins* story into *Lean-Sigma-TOC* language.

Our primary goal was to reduce deliver the same *value added work* to patients but with a shorter *lead time* by creating an outpatient *work-cell* that reduced *handoffs*, *time traps*, *queues*, *transport waste*, *motion waste and re-work*. The *current state map* was evolved in a series of planned *kaizen events* and the impact monitored using *run charts*. The communication *bottleneck* was resolved by replacing the *required work* of referral-by-letter with a shared electronic record. The *continuous quality improvement* monitoring revealed further *quality control* problems whose *root cause* was *demand variation* meeting a fixed clinic *capacity*. Our solution was to use *Pareto analysis* to create a *product matrix* that defined four *value stream product lines*. Using the *takt* time for each steam we created a *level schedule* at the rate of *customer pull* for each *value stream* to ensure low *latency* and high *utilisation* of the *pacemaker steps*. We used the *value stream map* to build a process simulation to conduct *resilience analysis* of a range of proposed the *future states*. The final design was shown to have low *sensitivity* to *common cause demand variation* and *special cause capacity variation*. The final design and was implemented as *standard work* using a *visual just-in-time booking template*. Further *kaizen events* have been completed that have reduced the *lead time* for generating clinic letters so that we complete *today's-work-today* and have further reduced the *transport waste* involved in moving the notes between clinic, office and medical records library. The reduction in *value stream cost* for the leg ulcer *value stream* has been greater than 26% through reduction in *over-processing*.

Three Wins

Chapter 8. The Change Engine

The three stages of change described in the previous chapters - Innovation, Investigation and Implementation appear to form a cycle. Innovation follows from implementation because implementation changes the external environment and this uncovers new problems; no sooner have you implemented one improvement than you reveal new problems and need to start the next cycle. There is no end – but on each turn the benefits, experience and confidence accumulate.

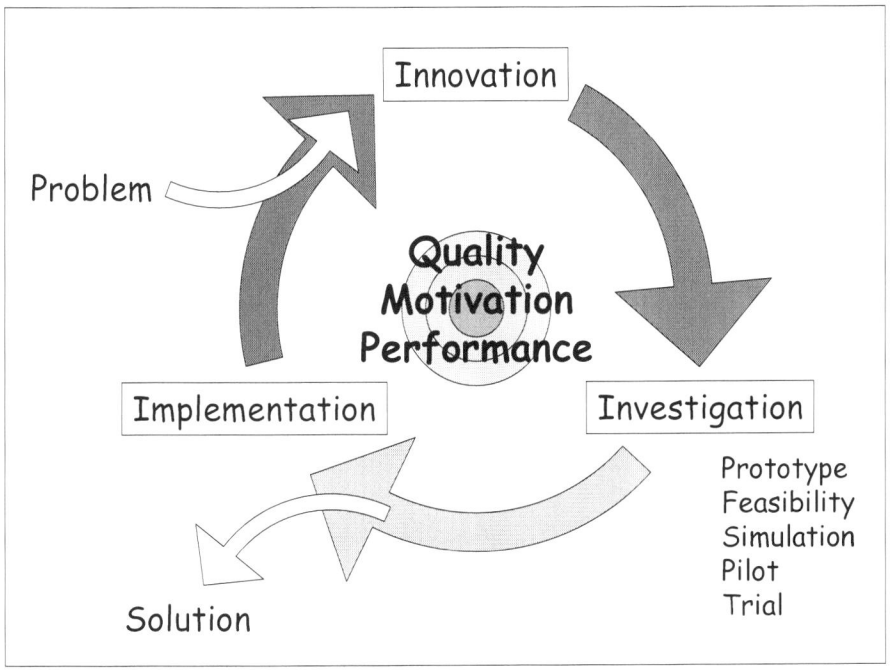

This concept of a change cycle is useful because each stage involves sub-projects that themselves are change cycles; wheels within wheels. All the wheels look the same, they just are just different sizes and turn at different speeds. Meredith Belbin described two further roles in successful teams that resonate with this concept of an endless cycle of change; *Shaper* and *Coordinator (previously called the Chair)*. The Shaper is the source of enthusiasm, drive and energy that provides the motive force for the wheel of change to turn; the Coordinator is more akin to the driver, the person who ensures that the wheel stays on course. The Shaper and Coordinator roles are important at all three stages of the project and although they have different roles they both need to maintain a strategic overview and not get lost in the detail. The concept of project or team leader appears to fit most closely with the Shaper and Coordinator roles.

The spiral

This concept of a change cycle is not new and the elements of it have all been described by many authors using different forms of language suited to different audiences. However, in the context of healthcare it has many guises: for example in the Scientific Method that underpins all medical research; innovation is represented by the hypothesis; investigation by the controlled experiment or trial; and implementation by the assimilation of new information into the existing body of knowledge. Similarly, in the Clinical Management model that underpins all medical training: innovation represents the formulation of the differential diagnosis; investigation represents the process of establishing the final diagnosis; and implementation represents the appropriate treatment plan for a specific patient. A more accurate metaphor is a *spiral* because as each cycle is completed you have not returned to the starting point; you have moved forward and outward; each cycle building on the previous one. The journey towards the win-win-win goal is not a straight line; it is spiral.

The lifecycle

All projects behave as if they have a life of their own; a predictable sequence of stages that can be labelled birth, growth, maturity, decline and death. Organisms and organisations demonstrate the same behaviour; it seems to be an inherent feature of all complex dynamic systems. Just as with the cycle of life there is generation of new cycles from existing ones, some close replicas of the original, and others very different. All innovation has a history, a family tree of ideas from which they are created. Occasionally a new mutation occurs, an invention or discover; a truly novel idea that often occurs by accident; but if the idea occurs in a nurturing context or confers a significant advantage, it will survive, be passed on, and will spread and appear in later generations. Change projects are no different - they are usually a mixture of many ideas from different places that have, by natural selection, been shown to be useful. One of the challenges that we face in continuous change is the incessant need to re-invent ourselves so by knowing how to surf the waves of change we can avoid getting swamped.

Our challenge ultimately required three cycles of change, each building on the previous one, each needing different techniques and tools to solve increasingly more difficult problems, and with the benefit of hindsight, each using the same principles to achieve success. The work has not finished - it will never finish - and having a deeper understanding of how we got this far is a good starting point for taking on new challenges. And these challenges continue to present themselves with unpredictable regularity.

> Our success with the LUTM system and the One Stop Clinic redesign has generated considerable interest from outside the organisation; which is what you would predict from the *Diffusion of Innovation* principle. The Resource Investigator / Early Adopters in other organisations have their ears, eyes and minds open and are prepared to try out ideas that have been proven to work. The LUTM system is already being used by early adopters at several other sites in the UK who have demonstrated the same benefits and rapid adoption. It has been interesting to observe the progress of these projects - some have moved quickly and some have moved slowly. The reason for the difference is not the validity of idea but the receptiveness of the organisation and their ability to implement change. That is the toughest challenge.

One reason for writing a book is to disseminate knowledge; not just of successful innovation but of successful implementation of innovation - apparently a much rarer event. At the beginning I stated that we did this with no project plan, no budget, no management input but this is the exception rather than the rule. The advice I would offer is use the right tools for each stage of the cycle; at the innovation stage keep free and unencumbered by protocol and procedure in order to nurture creativity; do not skip the investigation stage and use the appropriate rigorous methods to provide the evidence needed to identify the best option; the implementation stage is the place for steering groups, plans, budgets and the active management of the human dimensions of the transition.

The test

The test of our ability to manage change is evidence of completing a cycle of change; to take a specific problem and successfully implement a solution. To do this requires passing through the three stages in turn; stages that require different skills. With this insight we can gauge the competence of a person, team or organisation to manage change. As an exercise I have devised a quick diagnostic test of change-ability (see Appendix A).

Consider a specific problem that you know of. Ask yourself the following questions in order. If you reply "Yes" to a question you can move on to the next; if you reply "No" then you have finished.

Q1: Do you feel passionate about the problem?

Q2: Do you have any ideas how to solve it?

Q3: Do you have the skills to test and implement your ideas?

Your answers to these questions will provide an estimate of where you are on the Adopter Curve for your specific problem:

Q1 Passion	Q2 Ideas	Q3 Skills	Classification
N	?	?	Laggard
Y	N	?	Late Majority
Y	Y	N	Early Majority
Y	Y	Y	Early Adopter

Remember that we will "score" differently for different problems; some will engage our passion and curiosity, others will not. The same principle is true for everyone else – we are all different and yet we are all the same.

The win-win-win principle that has distilled from my own experience states clearly that the critical factor is the people; their shared vision; their individual motivation; their common knowledge; their individual experience and their unique perspectives. Not everyone needs to share this view; but enough do in order for the time and effort wasted on conflict to be re-directed into making progress towards a desirable and sustainable future.

Final thoughts

Writing this book has been as much a voyage of discovery as the story it relates. None of the ideas presented and discussed are new but the resonance and synergy between the different "theories" suggests an underlying principle of inter-dependence.

The Win-Win-Win metaphor seems to be a concise way of stating this principle and leads intuitively to the Motivation-Quality-Performance sequence that I believe encapsulates the route to successful management of change. I like to keep things simple and memorable which is why the "No Threats, No Mistakes, No Waste" is another way to express the Three Wins concept. An alternative stated in positive language is "Right Thing, Right Place, Right Time, First Time, Every Time". Which you prefer is just a matter of personal taste. That's OK.

Achieving a Three Wins outcome is about having a clear vision of what you want; it is about building relationships, inter-dependency, trust and co-operation; it is about respecting differences and combining strengths to bridge gaps; it is about creating something that is more than the sum of the parts; it is about seeing change as a positive opportunity to improve our own personal best; and it is a challenge that builds motivation and ultimately improves everyone's experience.

Each part of the journey has its own character, behaviour and methods; and each part of the journey is inter-dependent. By focussing on getting all the parts right - innovation, investigation and implementation - you build a spiral change stream that flows. As individuals we are naturally attracted to some parts of this process more than others; this is normal, expected and to be encouraged. In contributing to the whole we each work from our strengths and in the process learn to close the gaps in our knowledge and experience.

I have described first hand a real-life example of the Three Wins principle in action and illustrated the power of the paradigm. More than that I have shown how our spiral journey evolved in three phases with the use of increasingly powerful tools: PDSA, RCT and DES.

The principles I have demonstrated are not widely practiced in healthcare. Some of the solutions and tools were created from scratch because of lack of suitable alternatives, lack of resources or ignorance of what was needed or available. The DES tool is particularly powerful because it allows a complex process to be considered objectively from many viewpoints; helping to identify the root causes of problems in complex interconnected value systems, and allowing the creative energy of the team to be focussed on finding an effective solution.

Once these principles, methods and tools are mastered and used correctly, I believe anyone can conceive, plan and implement successful programmes of change for themselves, their teams and their organisations. Creating the synergy for a group of diverse stakeholders is made easier by establishing the common values, building a shared vision and proving a solution is designed to benefit everyone. Confidence and trust comes from increased knowledge and insight and removes the major barrier to change: the fear of failure.

I believe that anyone can achieve a Win-Win-Win future and each one of us can make that decision for ourselves. The journey starts with this single step; the second is to ask questions; the third is to listen to the replies.

Enjoy the journey.

References

Bagust A, Place M, Posnett JW. Dynamics of bed use in accommodating emergency admissions: stochastic simulation model. *Br Med J* 1999; **319:** 155-158.

Samad A, Hayes S, Dodds SR. Telemedicine: an innovative way of managing leg ulcer patients. *Br J Nursing* 2002; 11 (6 Suppl): S38-S52.

Dodds SR. Shared Community-Hospital care of Leg Ulcers using an Electronic Records and Telemedicine. *Int J Lower Extremity Wounds* 2002; **1:** 260-270.

Samad A, Hayes S, French L, Dodds SR. A comparative study of computerised digital image tracing versus contact tracing for objective measurement of leg ulcers. *J Wound Care* 2002; **11:** 137-140.

Hayes S, Dodds S. Telemedicine: a new model of care. *Nurs Times* 2003; **99:** 48-49.

Young T, Brailsford S, Connell Con, Davies R, Harper P, Klein JH. Using industrial processes to improve patient care. *Br Med J* 2004; **328:** 162-164.

Dodds SR. Designing improved healthcare processes using discrete event simulation. *Br J Health Comput Info Manag* 2005; **22:** 14-16.

Herzberg F, Mausner B, Snyderman BB. *The Motivation to Work.* Second Edition. John Wiley & Sons, New York. 1959.
ISBN:

Maslow AH. *Motivation and Personality*. Second Edition. Harper & Row, New York. 1970.
ISBN:

Covey SR. *The 7 Habits of Highly Effective People.* Simon & Schuster, London 1989.
ISBN: 0-684-85839-9

Johnson S. *Who moved my cheese?* Vermillion, London. 1998;
ISBN: 0-09181-697-1

Gladwell M. *The Tipping Point: How little things can make a big difference.* Abacus, London 2000;
ISBN: 0-349-11346-7

Fraser SW. *The Patient's Journey. Mapping, Analysing and Improving Healthcare Processes.* Kingsham Press, Chichester 2002.
ISBN: 1-904235-09-03

Rogers EM. *Diffusion of Innovations.* Fifth Edition. Free Press, London 2003.
ISBN: 0-7432-2209-1

Oldham Sir J. *Sic evenit ratio ut componitur - The small book about large system change.* Kingsham Press, Chichester 2004.
ISBN: 1-904235-27-1

Belbin RM. *Management Teams. Why they succeed or fail.* 2nd Edition. Elsevier, Oxford 2004
ISBN: 0-7506-5910-6

Peters T, Waterman RH. *In Search of Excellence.* Harper & Row, 2004.
ISBN: 1-86197-716-6

Gay T. *Simplicity is the key: Thoughts on leadership, change management and how things get done in organisations.* Kingsham Press, Chichester 2005.
ISBN: 1-904235-31-X

Bridges W. *Managing Transitions – Making the most of change.* 2nd Edition. Nicolas Brealey Publishing, London 2003.
ISBN: 1-85788-341-1

De Bono E. *Six Thinking Hats®.* Penguin 1999.
ISBN: 0-14-029666-2

Appendix A - The Innovation Questionnaire

This is a longer version of the Test summarised in Chapter 7 and is presented to illustrate how the ideas outlined in the book can be applied. If any answers are No then record the question number and stop.

Q1. Do you believe that the <system> can be improved?
Q2. Do you know why the <system> needs to be changed?
Q3. Do you have ideas of how to change the <system>?
Q4. Have you tried these ideas?
Q5. Was the system improved as a result?

Interpretation:
Q1: If you stopped here then you believe that the system is working as well as it can and you have no desire or need to change. You may be right and if everyone else who works with that system believes the same then you have no reason to change. If however, others do not agree then prepare to have some of your assumptions and your beliefs challenged - you maybe resisting change by being selectively blind to the signs that change is needed.

Q2: You feel the need for change but the cause of the problem is not obvious. Write down what are the symptoms that represent your feeling and keep asking yourself "Why?" to get to the root cause of each symptom.

Q3. You feel the need for change and think you know the cause of the problem but you don't know what to do. Follow the "*There is Nothing New under the Sun*" and "*Don't Re-invent the Wheel*" principles and start asking questions. Search inside yourself and look for others who have solved the problem; even in other disciplines. Find out how they did it.

Q4. You feel the need, you know the cause, you have a solution and you have good reason to believe it will work but you are being blocked. You need to identify the blocks and search for the least-resistance path ahead. In effect you need to demonstrate the skills of a leader to get to the next stage. There are really only two sources of blockages - people and resources. People may oppose your idea because they have good reason to challenge your assumptions - remember this is what you are doing so respect this right. This is actually a helpful form of opposition that if managed well can be used strengthen your case and can result in a new supporter: a win-win outcome. Other people may block your idea without offering any reason because is just easier to say "No". They are following their path of least resistance just as you are – it is just that they don't share your passion. This is a more difficult form of resistance to manage and generally it is better to go around such blocks than through them - canvas support from other more receptive people but leave the door open so that the "laggards" can get on board later if your idea proves to work - and be prepared to shrug off the "I told you so" comments if it doesn't - because they didn't tell you so.

Q5. If you answered "No" then you have successfully negotiated all the hurdles in the Change Cycle and had a chance to try your idea but it didn't result in what you wanted or the benefit was short lived. You may feel as though you have failed and the whole thing was not worth the effort and you will never try changing anything again. It fact you have succeeded but have not given yourself (or your team) due recognition. There are many reasons why things don't go as we planned: a plan that could not work but you didn't spot it; a plan that would work but the change process did not work; or a good plan, a good process but an unexpected change in the environment that moved the goal posts. In all these cases you still emerge stronger from the change process than you started - you always make some progress and learn in the process. You need to reflect about why you didn't reach your stated objective, learn from this, and "get back on the horse". Inexperience and enthusiasm often lead us to take on tasks that we are not yet competent to tackle and apparent failure dents our confidence. Develop the humility to seek out older and wiser counsel; listen to their arguments; and make your own mind up about what makes sense to you. It is a good strategy to break the problem up into smaller bits and solve the easy ones first, build your confidence and skills and then use these as tools on the more difficult problems - but keep an eye on the big picture. Too much confidence may be as counterproductive as not enough - it is said that ignorance and arrogance are a dangerous combination. The more confident of your ideas you are the more you are convinced you are right and everyone else is wrong the more you should test your ideas on others and listen to their arguments: they may have a view that you have not considered and you may be trying to force a win-lose outcome. This is "conflict thinking" – and the outcome will be the same as any conflict – both sides will lose overall. It is true that persistence is an important characteristic of an innovator but it needs to go hand in hand with open eyes, open ears and an open mind: seek out the active opposition - people who will explain why they do not agree with your ideas. If you find yourself in a conflict of opinion then you will need to get more understanding of the other person's point of view. It is more than likely you are in a state of "heated agreement" and you just need the evidence to make the common ground explicit. This is a very productive and constructive process if managed well. One characteristic of an effective leader is not to punish failure; to learn from mistakes; and not allow the same mistake to be made twice.

This test also re-emphasises the importance of motivation in the change process - without motivation nothing will happen – motivation means movement. It would be interesting to see how different people rank the three elements of the win-win-win principle: quality, motivation and performance. I think there will be a synergy between the *Win-Win-Win* principle and the *Diffusion of Innovation*, i.e. they are the same principle viewed in different ways. Steven Covey defines *principles* in just this way - Laws of Nature that exist independently of your opinion. I have an idea that the point at which motivation appears in the win-win-win ranking will also point to where you sit on the Adopter Curve

Win-win-win ranking	Classification
M►Q►P	Early Adopters (Quality driven)
M►P►Q	Early Adopters (Performance driven)
Q►M►P	Early Majority (Quality driven)
P►M►Q	Early Majority (Performance driven)
Q►P►M	Late Majority (Quality driven)
P►Q►M	Late Majority (Performance driven)

These two tests are simple to do and I believe help to manage a change project by engaging people at the stage of the project that suits them best. I have no hard evidence for this but it is an easily testable hypothesis.

Three Wins

Appendix B - The First Ten Steps

Do you want to rid your life of hassle? Do you want to feel happier and more secure? Do you want more wins? This short guide is designed to help you get started.

1. Focus on the Wins

2. Choose your Mindset

3. Try a New Perspective

4. Remove the Toxic Waste

5. One Step at a Time

6. Celebrate Progress

7. Invest to Save

8. Learn by Teaching

9. Ask for Help

10. Never Give Up.

Ask, listen, learn, practice, teach then ask again.

We all live in an uncertain and often frightening world so we each create a comfort zone to protect us. Our comfort zones are our physical and emotional nests; uniquely tailored to our needs and sculpted by our experience. We defend our comfort zones from threats; we fight to keep them as they are; we build strong walls around them and over time we make the walls higher, thicker and stronger to keep us safe.

The problem is that the world around us is always changing. We need to keep updating, expanding and even moving our comfort zones to keep them from being left behind by the tide of change. To do this we must look outside and see what is happening; and we must step out of our comfort zones to explore new possibilities.

Everyone is unique; no two people see the world the same way; there is no "one size fits all" solution that works for everyone. We each have to find our own path. This guide will ask questions. Some of these questions will be uncomfortable because they are designed to get you to think just outside your comfort zone. There are no right or wrong answers; you will know what feels right for you.

Are you ready to take the first step?

www.ThreeWins.com

1 Focus on the Wins

How good could your perfect future possibly be? What would your perfect future look like? Sound like? Feel like? Smell like? Taste like? Close your eyes and make the vision of your perfect future real in your mind.

Now ask yourself: "What is present in my perfect future that is not present in my life now?" Is it Confidence? Is it Optimism? Is it Achievement? Is it Happiness?

Happiness is a state of mind resulting from your interaction with the world around you. You can choose to feel happy; for a while. But being happy is only possible if your comfort zone is not threatened.

To achieve the Win-Win-Win goal you need to:

1. Build confidence by delivering your best to yourself and others;

2. Build optimism through motivating others by your example;

3. Build achievement by repeatedly exceeding your own personal best.

2 Choose your Mindset

Who or what is stopping you from being happy? Is someone or something else in your way? What do you do? Do you identify the target? Do you assess the threat? Do you choose your weapons? And when you are ready do you launch your attack? And what happens? Do you win or do you lose?

If this sounds familiar then you may be using a War metaphor. You may be seeing life as a perpetual series of battles to be won. So stop for a moment and ask "What is the cost of the battle for me and my opponent?" If you think about it the only answer is "It costs us both; we both lose!" The difference between the winner and loser of a battle is only how much you lose.

You are intelligent, capable and determined. If you really put your mind to it you will get what you set out to get. With a War mindset as your guide you will end up worse off than you started. Is that what you intended?

So, if you want to win you first need to choose a Win-Win-Win mindset.

Sounds simple enough; ready for the next step?

3 Try a New Perspective

Which is more rewarding - finishing the race in second place or not finishing at all? The guaranteed way to fail is to choose the "Can't Do" mindset and to not even try. With a "Can Do" mindset you at least have a chance of completing the journey to your win-win-win future. Make it a race. Picture yourself crossing the finish line. Hear the cheers. Feel the relief. Smell the sweat of effort and taste the tears of joy. Make the dream a vision.

Suppose you were your own customer? What value would you offer? Would you buy from yourself? Would you be disappointed? Would you come back for more? Would you recommend yourself to others? If not then why should anyone else value what you have to offer? The easiest way to see the value you offer is to become your own customer. Sounds simple enough, so how is it done? From now on, every time you disappoint yourself you must stop; acknowledge the feeling; hear the voice in your head; listen to what it is saying; look for the reason why; consider what you would do to prevent that feeling in future. Just picture you doing it better next time. Don't beat yourself up; become your own customer.

4 Remove the Toxic Waste

What gives you a really bad day? Is it what happens to you or is it how you react to what happens? What do you do when something irritates, annoys or niggles you? Do you accept it? Do you dismiss it? Do you fret about it? Do you complain about it? Or do you fix it?

Niggles are everywhere. Even if you try to ignore them they still effect you; insidiously, silently, subconsciously, persistently. Eventually you have to act; the Niggles are not your fault so you have to blame someone else. Niggles are like emotional toxic waste; and when you blame others you are creating and spreading more toxic emotional waste. And remember, you are living in the toxic emotional waste that you and everyone else created. Is that what you want?

Alternatively, when you experience a Niggle you can stop; acknowledge the feeling; step back; count to ten; look at the Niggle from all sides; uncover the root cause, and if you can, dig it out and put it in the bin forever. If you can't fix the Niggle immediately just mark it and park it. It has been uncovered so its days are numbered.

How does it that feel? Better? You bet it does!

5 One Step at a Time

How long is the journey to your Win-Win-Win goal? How do you know where you are and how much further there is to go? Look how many obstacles there are? This is impossible! Why even bother starting? If this is the way you see the journey then you are using a "Can't Do" mindset and are in danger of giving up before you start.

To achieve your goal all you have to do is believe that any step in the direction of your Win-Win-Win goal is worth the effort. Even if you can't see all the way; set your first goal within sight and make it possible to achieve. It is just one small step. And when you have taken it you will be able to see further than before. Just keep your next goal within sight; one step at a time.

If fact you have already completed several steps and you didn't really notice. You took the time to read this far. Ask yourself now "Have these steps taken me in the direction I want to go?" If your answer is "No" then stop; read no further; this is not the right path for you.

On the other hand, if your answer is "Yes" then keep going - and take the next step.

6 Celebrate Progress

How often does someone say to you "Well done!" How often does someone notice your contribution and go out of their way to thank you? How often do you do it for others? If you just thought "Not as often as I should" then stop; you just exposed a Niggle. Ask yourself "Why?" and keep asking "Why?" until you get to the root cause. If it is "Because I feel embarrassed to give praise" then relax, that is what most people say.

Now ask yourself what you do when someone gives you genuine recognition? Do you ever dismiss it? How does that feel for the person giving you the praise? Could that be why you are reluctant to give praise - because someone might dismiss it and therefore discount you?

Not being able to accept praise is the root cause of not being able to give it. Try this exercise now. Practice what you would say the next time someone gives you praise. Picture you with a big smile on your face saying "Thank you, very much. You just made my day".

Then next time someone recognises your contribution you'll be ready to accept it and to feel better for it.

7 Invest to Save

How do you feel when you see something that needs to be done and all you hear is "That's not my job!" Is that a Niggle? If so just ask "Why?" Are the people who say this lazy, incompetent, or have they just learned a "Can't Do" habit and no longer notice when they are being poisoned by their own toxic emotional waste.

What will you do now? Ignore it? Get angry? Complain? Do it yourself? Or do you expose the Niggle? You have the opportunity to help someone else learn how to clean up toxic emotional waste; and that's good for everyone. Ask them "Why?" and listen to their answer. Try to put yourself in their shoes and understand why they gave you that answer. Ask yourself what would have to happen for them to say "I'd be happy to help." If you can help them learn to recognise and dispose of Niggles too then you are investing in your own toxic-waste-free future.

Remember: that no investments come with the guarantee of a return; when a return might happen; or how large the return might be. You can be sure of one thing though:

If you never invest you will never get a return.

8 Learn by Teaching

How do you learn? Is it by reading books; by reflecting on your experiences; by trial and error; or by asking for advice? Remember: if you want to learn you need to be prepared to ask for help. So consider for a moment those that you ask advice from. What are they trying to achieve? How will they feel if you don't learn? How can you help them to help you? If you can see the common ground where you both achieve what you want then you have found part of a journey to your win-win-win futures that you can share.

Being the teacher; finding the common ground and learning how to teach is part of learning how to learn. It is a good way to practice your skills. It is a long term investment in your own win-win-win future.

Teaching how to learn is a better long-term investment than teaching how to do. Demonstrating to others to choose a Win-Win-Win mindset; to be aware of others; and to dispose of niggles is a secure investment in your future success.

To teach you must learn; to learn you must change; to change you must step outside your comfort zone.

9 Ask for Help

What do you advise others to do if they don't know what to do? What happens when you don't know what to do? Do you take your own advice? Do you walk your own talk? The quickest way to find an answer is to ask someone who knows it. You do not need to know everything; you only need to know who to ask.

The Three Wins metaphor is linked to a simple concept called a Value Stream. Value is what you get when you reach the next goal on your journey. To reach your Win-Win-Win goal you must create a personal value stream that flows; the faster the flow the quicker you get there. It does not matter if your job is to design new products, to make those products, or to support the customers who buy those products. It does not matter if your purpose is to create personal wealth, wisdom, happiness or all three. Any path to your Win-Win-Win goal is a value stream.

So, start by flushing your personal value stream of the toxic emotional waste that is poisoning it. Clean out the Niggles that clog up the stream and sap your emotional energy. Only then will you have the time and energy to invest in learning, in changing and in moving forward.

10 Never Give Up

What do you do when you can't have something you really want, have worked for and feel you deserve? Do you give up or do you keep trying? Not getting a fair reward is a Niggle. The path to your Win-Win-Win goal is the Niggle-free path of fair rewards. How long will it take? The future is not yet written and the past cannot be unwritten. You are in the hands of the present and the only influence you have on the future are the choices that you make now.

The journey starts easily enough; you only have to choose to want to do it. And you must commit to keep making that choice. When you give up wanting to beat you own personal best you will stop moving forward.

Learning to make choices that take you towards your Win-Win-Win goal is the challenge. Learning takes time, learning takes practice and if you want progress quickly you will not have time to learn by trial-and-error. Anyone who is farther along the path can help you because they can see what is ahead that you can't.

So, when you are ready, make your choice. Commit to the challenge. And never give up.